Internet of Things (IoT) Certification Program for Game Changers

Learn. Practice. Get Certified.

ANGEL BERNIZ

International Institute of Executive Careers (IIEC)

Program Contents

Welcome to the Program

Thank you very much for purchasing this book.

First of all, let's connect on LinkedIn and join my **Game Changers** Newsletter on LinkedIn, so that you can get my insights, learnings, and exclusive contents to help you get career success:

https://www.linkedin.com/in/angelberniz

I would love you to share your experience with me after completing the program. You can message me via LinkedIn.

Okay, let's move on!

Certification Program

This book is the foundation for the **IIEC IoT Certification Program™** available here:

https://thebestmba.org/program/iot-certification

The Certification Program is available in the online program and will improve your job opportunities and certify your skills gained from this book.

The exams in the online Certification Program are based on the exam questions covered along this book. And once you sign up, you can take your online program whenever you want (no deadline limits). Just sign up today for your Certification Program and certify your skills.

Now, let's start enjoying this journey.

Preface

As we progress through the 21st century, we find ourselves at a crossroads, where rapid advancements in the Internet of Things (IoT) and associated technologies are dramatically redefining our industries and everyday lives. The IoT era ushers in both unparalleled challenges and transformative opportunities for those who aspire to pioneer this change and make an enduring impact. It's becoming increasingly essential for these change-makers to adopt a practical, secure, and inclusive approach to navigating this intricate and evolving terrain.

The aim of this program is to serve as a practical and accessible resource for these change-makers, enabling them to unlock the potential of IoT and lead the way in shaping our connected future. Drawing insights from a diverse array of fields, including computer science, engineering, social sciences, design, and security, this program seeks to foster a comprehensive and interdisciplinary understanding of the IoT era and its implications for our increasingly interconnected world.

Throughout the program, we will establish a foundation for understanding the key principles and concepts that underpin IoT technologies and discuss the security considerations that emerge as we integrate these advancements into various sectors. The program offers strategies, best practices, and actionable insights designed to empower change-makers in their pursuit of innovation and positive transformation.

As we explore the future in the IoT era, we'll examine the potential impact of IoT technologies on sectors such as manufacturing, healthcare, transportation, and homes, highlighting both the transformative opportunities and the challenges that lie ahead.

Throughout the program, we'll emphasize the need for responsible innovation, interdisciplinary collaboration, public engagement, and global cooperation to navigate the complexities and seize the opportunities presented by the IoT era. By actively participating in shaping our connected future, change-makers can ensure that IoT technologies are developed and deployed in ways that align with our collective values, security principles, and societal goals.

By offering this program, I hope that participants will gain a deeper understanding of the IoT era and its implications for our interconnected world, as well as a sense of optimism, inspiration, and empowerment. As change-makers, you have a unique opportunity to shape the trajectory of IoT development in ways that reflect our shared aspirations for a more efficient, secure, and sustainable world. Together, let's embrace the challenges and opportunities of the IoT era and chart a bright future for humanity in this interconnected world.

Acknowledgments

My notes at the university were: "Early adopters are the wealthiest people on earth."

Little did I know how those words would shape my path. As I reflect on my journey, I am overwhelmed with gratitude for the countless individuals who accompanied, guided, and inspired me along the way.

I owe a profound debt of gratitude to my mentors, colleagues, and collaborators in executive education and the technology industry. Their unwavering support and willingness to share their knowledge and perspectives have been invaluable. Their guidance has shaped my career and fueled my passion.

To my students at the International Institute of Executive Careers (IIEC), you have been my constant inspiration. Your curiosity and dedication have made me a better educator and challenged me to refine my ideas. Your insights have significantly contributed to this book.

I also acknowledge the remarkable researchers, professionals, and practitioners who have paved the way for our understanding of the potential of emerging technologies. Their visionary thinking and tireless pursuit of knowledge continue to inspire individuals worldwide, including myself.

I am deeply grateful for the unwavering love, support, and understanding of my friends and family. Your belief in me and willingness to engage with my ideas have given me strength and resilience. This book is dedicated to you with love and gratitude.

As I conclude, I am reminded of the interconnected relationships, experiences, and ideas that have shaped this book. To all who have touched my life, big or small, thank you. Your presence has been truly meaningful. Together, let us continue exploring the evolving landscape of technology, guided by the belief that early adopters can shape the future.

About the Author

Angel Berniz is a seasoned Business and Technology executive, strategist, and advisor with over 20 years of experience in the technology industry. He has held various leadership positions in several companies and has worked with clients in different industries, helping them navigate cloud transformation, innovation, sustainability, emerging technologies, and cybersecurity.

Angel Berniz is also a recognized keynote speaker and author on topics related to business technologies, leadership, and innovation. He has been featured in several conferences, seminars, and publications. He is also a professor and lecturer in international executive programs of several Business Schools.

You can explore Angel Berniz's extensive collection of books available on Amazon:

https://www.amazon.com/author/angelberniz

Module 1. Introduction to Internet of Things (IoT)

Introduction to IoT

Welcome to Module 1on Introduction to Internet of Things (IoT). In this module, you will embark on an exciting journey into the world of IoT, a realm where physical and digital worlds intersect, creating a wave of transformation that is redefining our lives and businesses.

The Internet of Things, often abbreviated as IoT, is an evolving field that has grown immensely in the last decade. It's a dynamic, interconnected system of devices, sensors, and software that share data over the internet. From the smart thermostat in your home that learns and adjusts to your routines, to the wearable fitness tracker on your wrist that monitors your health, and to the complex systems in industries that optimize operations, IoT is everywhere.

But why is this important? The power of IoT lies in the value of connected data. The ability to capture, analyze, and make informed decisions from this data is revolutionizing industries, creating more efficient processes, enhancing productivity, reducing costs, and even sparking new business models. The impact is so profound that it's often compared to the advent of electricity or the internet itself.

Consider this practical example: A farmer uses IoT sensors to monitor soil moisture levels, weather conditions, and crop health in real-time. This data can be analyzed and used to automate irrigation, optimize fertilizer use, and predict the best times for planting and harvesting. This is not a scene from a sci-fi movie, but a real-life application of IoT technology that is transforming agriculture today.

The journey through IoT is not just about understanding its technical aspects, but also recognizing its broader implications on society, economy, and our daily lives. As the world becomes increasingly digitized and interconnected, it is essential to understand the principles of IoT, its potential applications, and the challenges it poses.

Over the course of this module, we will unpack the concept of IoT, trace its evolution, explore its key components and architecture, and delve into a variety of its practical applications and use cases. Whether you're a student, a professional, an entrepreneur, or simply a curious mind, this module is designed to provide a comprehensive introduction to IoT, setting a solid foundation for your journey into this fascinating field.

Let's get started on this exciting journey to demystify the Internet of Things!

What is IoT?

At its core, the Internet of Things, or IoT, is a concept that refers to the interconnection of everyday objects or devices to the internet, enabling them to send and receive data. This might seem like a simple concept, but its implications and potential applications are vast and complex.

Imagine waking up in the morning to the sound of your alarm clock. But this is not just any alarm clock. It's a smart alarm clock, connected to the internet. It not only wakes you up, but it also communicates with your coffee maker, signaling it to start brewing your morning coffee. Meanwhile, your smart thermostat adjusts the temperature of your home based on the time and weather, ensuring it's just right when you get out of bed. This is a small, yet practical, glimpse into your personal life empowered by IoT.

But IoT extends far beyond smart homes. Now picture a city where the street lights are interconnected through the internet. These smart lights can adjust their brightness based on the time of day or weather conditions, conserving energy when full brightness isn't needed. They can also relay information about outages immediately, allowing for faster repairs. This is a basic example of a smart city, another domain where IoT is making significant impacts.

So, you might wonder, how does IoT work? Each 'thing' in IoT is equipped with sensors or devices that can collect and transmit data. These devices are often connected to an IoT platform that collects, processes, and analyzes this data, making sense of it. This data can then be used to make informed decisions or trigger actions.

Let's consider a practical example to illustrate this. Imagine you have a smartwatch that monitors your heart rate. The watch, an IoT device, collects data about your heart rate throughout the day. This data is sent to an IoT platform where it is analyzed. If your heart rate spikes or drops abnormally, the platform can send a notification to your doctor or even trigger a call to emergency services. This is IoT at work, potentially saving lives.

However, IoT is not without its challenges. Issues such as data security, privacy, and interoperability are significant concerns that need to be addressed as you delve deeper into this field. Despite these challenges, the opportunities and benefits that IoT brings to various aspects of our lives are vast and transformative.

At this point, you may be thinking, "Why should I care about IoT?" The answer is simple. As digital interconnectivity continues to pervade every facet of our lives and work, understanding IoT will help you navigate this evolving landscape. Whether you wish to leverage IoT technologies in your business, build IoT solutions, or simply make informed decisions as a consumer, a clear understanding of what IoT is, is your first step in this journey.

The depth and breadth of IoT make it a complex field to fully grasp in one go, but don't worry. As we move forward in this module, we'll be breaking down these concepts further, exploring the technologies that enable IoT, its various applications, and the challenges it poses. As you immerse yourself in this learning journey, you'll find that IoT is not just a technological revolution; it's a catalyst for transforming how we live, work, and interact with the world around us.

Evolution of IoT

To truly understand the Internet of Things (IoT), it's important to take a step back and explore how it evolved over time. The journey of IoT is a testament to human ingenuity, weaving together strands of technology, connectivity, and data into a vast, interconnected network that has transformed the world as we know it.

The roots of IoT can be traced back to the 1980s and 1990s, at the dawn of the internet age. However, the term "Internet of Things" was not coined until 1999, by Kevin Ashton, a British technology pioneer. Despite this, the concept of interconnected devices had been percolating for years. The first recognizable IoT device was a modified Coca Cola vending machine at Carnegie Mellon University in the early 1980s. This machine was reprogrammed to report its inventory and whether the newly loaded drinks were cold, a seemingly simple innovation that marked the beginning of a new technological era.

As the internet evolved and expanded in the 1990s and early 2000s, so did the idea of connected devices. The advent of wireless communication technologies, such as Wi-Fi and Bluetooth, provided the necessary infrastructure for devices to communicate without physical connections. This was a crucial step in the development of IoT, as it allowed for a much greater range and flexibility of connected devices.

Fast forward to the late 2000s and 2010s, the era when IoT truly began to take shape. In 2008, the number of devices connected to the internet surpassed the number of people on Earth, a monumental milestone in the evolution of IoT. The following year, the term "Internet of Things" was added to the Oxford English Dictionary, signifying its growing prominence.

By this time, advancements in sensor technology, cloud computing, and data analytics had paved the way for more sophisticated IoT applications. The cost of sensors and processors had decreased, making it feasible to integrate them into everyday objects. Meanwhile, cloud platforms had emerged as a scalable solution for storing and processing the vast amounts of data generated by these devices. This convergence of technological advancements was a catalyst for IoT, ushering in a new wave of applications across various sectors.

Consider the evolution of the automotive industry, for instance. Early stages of IoT in cars were largely centered around navigation and entertainment systems. Today, with the advancements in IoT, your car is not just a mode of transportation. It's a hub of interconnected systems, capable of providing real-time diagnostics, tracking fuel efficiency, offering predictive maintenance, and even enabling autonomous driving.

Now, in the 2020s, IoT is at the forefront of the Fourth Industrial Revolution, also known as Industry 4.0. It's propelling advancements in sectors like manufacturing, healthcare, agriculture, and urban development, among others. The advent of technologies like 5G, edge computing, and Artificial Intelligence (AI) is driving the next phase of IoT evolution, pushing the boundaries of what's possible.

The evolution of IoT is a continuous journey. As you navigate through the history of IoT, you not only gain insights into its transformative impact over the years but also glean valuable lessons from its evolution. Understanding the past and present of IoT can help shape your perspective on its future potential and identify opportunities for innovation, improvement, and impact.

Remember, IoT is not just a technological transformation; it's a change in mindset. It's about seeing connections, creating value from data, and leveraging technology to solve problems and enhance lives. As we dive deeper into the intricacies of IoT in the upcoming sections, you'll uncover the multi-faceted nature of this field, revealing not just its technological dimensions, but also its societal, economic, and ethical implications. The evolution of IoT is far from over, and you are part of this ongoing narrative. As you explore the past milestones and current state of IoT, you also get a preview of the exciting possibilities that lie ahead.

In the realm of healthcare, for instance, IoT is not just about wearable fitness trackers anymore. We are moving towards smart hospitals with connected medical devices that monitor patient vitals and send alerts for any anomalies, telemedicine platforms that enable remote consultations, and even IoT-enabled medical implants that can deliver medication within the body. You may one day find yourself at the center of a personalized healthcare system powered by IoT, where proactive and predictive care is the norm.

In the domain of smart homes, we've gone beyond connected light bulbs and thermostats. We're now looking at integrated smart home systems where your refrigerator can track your food inventory and place grocery orders, your washer and dryer can optimize cycles based on the weather forecast, and your security system can recognize faces. You could be living in a home that not just automates tasks but also anticipates your needs and learns from your habits.

In the industrial sector, the evolution of IoT has led to the rise of the Industrial Internet of Things (IIoT). IIoT involves the use of IoT technologies in manufacturing and industrial practices, making them more efficient, safe, and cost-effective. It's not just about automating processes but optimizing entire supply chains, preventing equipment failures, and even enabling new business models like product-as-a-service. Your role in the industry might evolve as IIoT becomes more prevalent, requiring a new set of skills and knowledge.

The ongoing evolution of IoT is also driving significant changes in cities and public spaces. The concept of smart cities is becoming a reality with connected infrastructure, public services, transportation, and even waste management. Imagine a city where traffic flow is optimized in real-time, public utilities are managed efficiently, and safety is enhanced through smart surveillance – this could be your city, transformed by IoT.

Understanding the evolution of IoT is like looking into a rear-view mirror while driving forward. It gives you a clearer view of where you've come from, provides context for where you are now, and guides you towards where you're headed. But remember, you're not just a passenger on this journey; you're also the driver. You can influence the direction and speed of IoT's evolution through your actions, innovations, and decisions.

In the upcoming sections, we'll dig deeper into the components and architecture of IoT, explore the enabling technologies, and delve into various applications across different sectors. As you continue this journey, you'll realize that the evolution of IoT is not just a timeline of technological advancements but a story of human progress and transformation. And it's a story that's still being written, with many more exciting chapters to unfold.

IoT Components and Architecture

To truly comprehend how the Internet of Things (IoT) operates and the range of possibilities it brings, it's essential to understand its key components and the underlying architecture. The complexity of IoT lies in its intricacy, a network of interwoven elements working in unison to capture, transfer, and interpret data, ultimately driving actionable insights and automated responses.

At a high level, an IoT system consists of four primary components: Sensors/Devices, Connectivity, Data Processing, and User Interface. Each of these components plays a critical role in the overall function of the IoT system, and understanding their operations and interactions provides a comprehensive view of how IoT works.

Sensors/Devices

Sensors and devices are the eyes and ears of an IoT system. They capture information from the environment, converting physical parameters into digital data. This could be anything from temperature readings, light levels, motion detection, to more complex data like images from a camera or sound from a microphone.

For instance, in your smart home setup, you might have a variety of IoT devices like a smart thermostat that measures temperature, a connected camera that captures images, and a smart lock that detects if a door is open or closed. Each of these devices uses different sensors to capture specific types of data.

Connectivity

Once the data is collected, it needs to be sent to a place where it can be processed and analyzed. This is where connectivity comes in. Depending on the type and volume of data, as well as the power and range requirements, different connectivity options may be used. These can range from cellular and Wi-Fi connections to more specialized IoT connectivity options like LoRaWAN, Sigfox, or Narrowband IoT (NB-IoT).

Using the smart home example, your smart thermostat might use Wi-Fi to send temperature data to a cloud platform, while a soil moisture sensor in your garden might use a low-power wide-area network (LPWAN) technology like LoRaWAN due to its lower power consumption and longer range.

Data Processing

Once the data reaches its destination, it needs to be processed and analyzed. This could happen on a cloud platform or on the edge, closer to where the data is generated. Data processing involves cleaning, validating, and analyzing the data to extract actionable insights.

Let's say, in your smart home, the temperature data from your thermostat is analyzed in real-time on a cloud platform. If the temperature falls below a certain threshold, it might trigger an action like turning on the heating system.

User Interface

The last key component of an IoT system is the user interface. This is how the insights and actions derived from the data are communicated back to you. This could be through an app on your smartphone, an email notification, or even a visual display.

In your smart home, if the soil moisture sensor in your garden detects that the soil is too dry, you might receive a notification on your phone suggesting you to water the plants. Or, if your smart lock detects that your door is open while you're away, it could send you an alert.

Understanding these components and how they fit together forms the basis of the IoT architecture. However, it's important to note that real-world IoT systems can be much more complex. They might involve multiple sensors and devices, different types of connectivity, and complex data processing algorithms. They might also need to consider aspects like security, privacy, device management, and interoperability.

As you delve deeper into the world of IoT, you'll start to see how these components and architecture come to life in different applications and scenarios. Whether it's a smart home, a smart city, or an industrial IoT setup, the same fundamental principles apply. And as you understand these fundamentals, you'll be better equipped toapply this knowledge across a range of contexts, be it designing your own IoT solutions, evaluating the IoT readiness of various products or services, or even anticipating the future trajectory of IoT technology. With a firm grasp of IoT components and architecture, you can navigate the technological landscape with confidence, making informed decisions and innovative contributions. Whether you are troubleshooting a complex IoT system, developing IoT strategies for a business, or just enhancing your personal tech environment, your understanding of IoT's underlying principles will be the key to your success.

IoT Applications and Use Cases

The true power of the Internet of Things (IoT) comes to life when you begin exploring its applications and use cases. From the comfort of your home to the hustle of cities, from healthcare to agriculture, IoT is transforming the way you interact with the world around you. Let's dive into some of these practical applications, where the convergence of sensors, connectivity, data processing, and user interface creates innovative solutions and opportunities.

Smart Homes

Starting with your immediate surroundings, smart homes are one of the most common applications of IoT. Here, various devices and appliances in your home are connected to the internet, enabling you to monitor and control them remotely.

Imagine you're on your way home from a long day at work. As you approach your home, your smart garage opener detects your car and automatically opens the garage door. As you step into your home, your smart lights turn on, and your smart thermostat adjusts the temperature to your preferred setting. Later, when you're ready for bed, a single command to your smart speaker can turn off all the lights, lock the doors, and set your alarm for the next morning.

Wearables

In the realm of personal technology, wearable devices have gained popularity, providing a wealth of information right at your fingertips, or more accurately, on your wrist. Fitness trackers and smartwatches can monitor your heart rate, track your steps, measure your sleep quality, and even detect if you've fallen and can't get up. The data collected by these devices can help you maintain a healthier lifestyle and could even save your life in an emergency.

Smart Cities

Zooming out from your home to your city, IoT is playing a crucial role in developing smart cities. In a smart city, various public and private sectors collaborate using IoT to improve efficiency, sustainability, and the quality of life for its residents.

Picture a city where connected sensors monitor air quality in real-time, providing you with updates so you can plan your outdoor activities. Smart traffic lights adjust to real-time traffic conditions, reducing your commute time. Connected waste bins notify city services when they're full, ensuring timely waste collection and a cleaner city.

Healthcare

In the field of healthcare, IoT is revolutionizing patient care and medical services. IoT devices can monitor patient vitals in real-time, alerting healthcare providers of any critical changes, no matter where the patient is located.

Imagine a future where your wearable device continuously monitors your health parameters, and in case of any anomalies, immediately shares this data with your healthcare provider. If required, medication is dispatched to your location, or an ambulance is called, potentially saving precious time during emergencies.

Agriculture

Even in agriculture, IoT is making significant strides. Precision farming, powered by IoT, optimizes the use of resources like water, fertilizer, and seeds, leading to increased crop yields and sustainability.

Consider a farm where soil moisture sensors ensure crops receive just the right amount of water. Drones monitor crop health from the sky, detecting any signs of disease or stress. When it's time to harvest, autonomous tractors do the job, guided by GPS and real-time field data.

Industrial IoT

In industries, the Industrial Internet of Things (IIoT) is revolutionizing manufacturing processes, supply chain management, and predictive maintenance.

Envision a factory where connected machines on the production line communicate with each other, optimizing production in real-time. Sensors on equipment predict potential failures before they occur, minimizing downtime. Even the inventory management is automated, with real-time tracking of raw materials and finished products.

Energy

In the energy sector, IoT is not just improving efficiency but also promoting the use of renewable energy. Smart grids are using IoT to optimize the generation, distribution, and consumption of electricity.

Think about your home where a smart meter not just measures your electricity consumption but also helps you understand and manage your usage patterns. On a larger scale, IoT-enabled wind turbines and solar panels can adjust their operations based on real-time weather data, maximizing their output. Electric vehicle charging stations can balance the demand and supply of electricity, ensuring optimal usage and promoting the use of electric vehicles.

Retail

In retail, IoT is enhancing both the customer experience and the retailer's operations. From smart shelves to automated checkouts, IoT is reshaping the retail landscape.

Imagine walking into a store where a connected shopping cart recommends products based on your shopping history. As you move through the store, digital signs display personalized deals and offers. When you're done shopping, you just walk out of the store, and the payment is automatically processed through your mobile app, eliminating the need for traditional checkouts.

Transportation and Logistics

IoT is playing a pivotal role in the transformation of transportation and logistics. Connected vehicles, smart logistics, and intelligent traffic management systems are improving safety, efficiency, and sustainability.

Consider a future where your car is connected to a network of other cars, traffic lights, and sensors on the road, providing real-time information to optimize your route, avoid traffic, and prevent accidents. In logistics, IoT-enabled fleet management systems can track and monitor vehicles in real-time, improving the efficiency and reliability of deliveries.

Environment

Lastly, IoT is contributing significantly to environmental monitoring and conservation. IoT devices can monitor air and water quality, track wildlife, and even predict natural disasters.

Envision a world where connected sensors in the ocean provide early warnings for tsunamis, giving people more time to reach safety. IoT devices can also track the health and movement of endangered species, providing valuable data for conservation efforts.

Education

In the field of education, IoT is paving the way for smarter and more effective ways of learning. From connected classrooms to personalized learning experiences, IoT is reshaping education.

Imagine an IoT-enabled education system where each student's learning style and pace is understood, and personalized content is delivered accordingly. Attendance is automatically recorded as students enter the classroom, and teachers can track each student's progress in real-time, allowing them to provide timely and targeted feedback.

Hospitality

In the hospitality industry, IoT is elevating guest experiences while improving operational efficiencies. Hotels are using IoT to provide personalized experiences to their guests while optimizing energy usage and staff workflows.

Imagine staying in a hotel where the room adjusts its lighting, temperature, and even the music based on your preferences as soon as you check in. The minibar automatically notifies the staff when it needs to be refilled, and the management gets real-time insights about energy consumption, allowing them to optimize their operations.

Construction

In the construction industry, IoT is enhancing safety, productivity, and efficiency. Connected equipment, wearable tech, and smart materials are some of the ways IoT is revolutionizing construction.

Consider a construction site where workers are equipped with wearables that monitor their health and safety in real-time. Machines and equipment are connected, providing real-time data about their operation and maintenance needs. Even the construction materials are smart, capable of sensing and adapting to their environment.

Insurance

In the insurance sector, IoT is transforming the way risk is assessed and premiums are calculated. From telematics in auto insurance to connected health devices in health insurance, IoT is enabling more personalized and fair insurance models.

Imagine an insurance model where your car insurance premium is based on your driving behavior, recorded by a connected device in your car. Your health insurance premium is influenced by your lifestyle and health parameters, tracked by your wearable device.

Media and Entertainment

In the media and entertainment industry, IoT is creating immersive and interactive experiences. Connected devices are not just changing the way content is consumed but also how it's created and distributed.

Consider a future where you can watch a movie or play a game in a fully immersive environment, interacting with the content in real-time. Your preferences and interactions are captured by connected devices, providing personalized recommendations and experiences.

The applications of IoT extend to virtually every industry and sector. They illustrate how IoT is not just about connecting "things" to the internet but about transforming businesses, industries, and lives. As you continue your journey into the world of IoT, you'll uncover more such applications, each offering new opportunities and challenges. They will inspire you, challenge you, and guide you as you navigate the exciting world of IoT.

Module 1 Exam Questions

1. What is the Internet of Things (IoT)?

a) A network of connected devices

b) A type of software

c) A new programming language

d) A type of computer

2. When did the term "Internet of Things" first emerge?

a) Late 1990s

b) Early 2000s

c) 2010s

d) 1980s

3. Which of the following is a key component of IoT architecture?

a) Connectivity

b) Data Processing

c) User Interface

d) All of the above

4. What is a smart home?

a) A house built with intelligent materials

b) A home where devices and appliances are connected to the internet

c) A home with high-speed internet

d) A home with a high IQ

5. What is a common use of IoT in agriculture?

a) Soil moisture sensors

b) Automated milking machines

c) GPS-guided tractors

d. All of the above

6. What is the role of IoT in smart cities?

a) Monitor air quality

b) Control traffic lights

c) Notify when waste bins are full

d) All of the above

7. What is the Industrial Internet of Things (IIoT)?

a) Use of IoT technologies in manufacturing and industrial processes

b) An internet connection for industrial use

c) A type of industrial machinery

d) A network of industrial computers

8. How does IoT benefit the energy sector?

a) By creating more fossil fuels

b) By optimizing the generation, distribution, and consumption of electricity

c) By reducing the need for electricity

d) By providing free electricity

9. How is IoT used in the retail sector?

a) To provide personalized deals and offers

b) To automate checkouts

c) To monitor inventory in real-time

d) All of the above

10. What's a popular personal application of IoT technology?

a) Smart glasses

b) Fitness trackers

c) Smart shoes

d) None of the above

11. How is IoT used in the transportation sector?

a) To optimize route and prevent accidents

b) To improve fuel efficiency

c) To enhance driver comfort

d) All of the above

12. How does IoT contribute to environmental monitoring?

a) By predicting natural disasters

b) By tracking wildlife

c) By monitoring air and water quality

d) All of the above

13. What is one way IoT is transforming education?

a) By providing personalized learning experiences

b) By eliminating the need for teachers

c) By making textbooks obsolete

d) None of the above

14. How does IoT enhance guest experiences in the hospitality industry?

a) By providing personalized room settings

b) By serving guests faster

c) By offering online check-in and check-out

d) All of the above

15. How is IoT used in the construction industry?

a) To enhance safety and productivity

b) To replace human labor

c) To design buildings

d) None of the above

16. How is IoT impacting the insurance sector?

a) By changing the way risk is assessed

b) By lowering insurance premiums

c) By eliminating insurance fraud

d) None of the above

17. How is IoT creating immersive experiences in media and entertainment?

a) By offering 3D movies

b) By providing interactive gaming experiences

c) By personalizing content recommendations

d) All of the above

18. Which of the following is a potential future application of IoT in healthcare?

a) Automated insulin delivery for diabetics

b) Smart pills that provide real-time health data

c) Remote monitoring of patient health

d) All of the above

19. How is IoT used in smart logistics?

a) To track and monitor vehicle locations in real-time

b) To control traffic

c) To facilitate automated checkouts

d) None of the above

20. How is IoT improving operations in the hospitality industry?

a) By automating room cleaning

b) By optimizing energy usage

c) By providing personalized guest experiences

d) Both b and c

Answers:

1. a
2. a
3. d
4. b
5. d
6. d
7. a
8. b
9. d
10. b
11. d
12. d
13. a
14. d
15. a
16. a
17. d
18. d
19. a
20. d

Module 2. IoT Technologies and Protocols

Introduction to IoT Technologies and Protocols

Welcome to Module 2 on IoT Technologies and Protocols. In this module, you will delve into the integral mechanisms that make the Internet of Things (IoT) a reality. You will learn about the foundational technologies and protocols that connect the physical world to the digital realm, enabling a seamless exchange of data across devices, systems, and services.

The IoT ecosystem is built on a myriad of technologies, each catering to unique requirements of connectivity, range, power consumption, and data transmission rates. Understanding these technologies and protocols is not only crucial for grasping the operational essence of IoT but also for appreciating the complexity and diversity of this transformative phenomenon.

You will be introduced to various communication technologies that provide the backbone for IoT connectivity. These range from traditional wired connections to the latest wireless technologies that are reshaping the landscape of IoT. With the proliferation of connected devices, each with its own unique communication requirements, the role of these technologies cannot be overstated.

Next, you will explore the different network protocols that govern how devices talk to each other in the IoT world. These protocols ensure that data flows smoothly and securely between devices, regardless of their manufacturer or the nature of the network they are connected to.

Further, you will uncover the data and message protocols that dictate the structure and format of the data being transferred. These protocols ensure that the data being sent from a sensor in a remote field can be understood by a cloud-based application thousands of miles away.

Finally, you will learn about device management protocols. As the number of connected devices scales up, managing them — from their configuration to their lifecycle — becomes a significant challenge. These protocols provide a systematic way to control and manage the IoT devices.

Throughout this module, we will focus not just on the theoretical aspects but also on practical examples. We will take you through real-life cases where these technologies and protocols are being employed. You will see how they come together to facilitate applications like smart homes, autonomous vehicles, precision agriculture, and more.

By the end of this module, you should have a solid understanding of the various technologies and protocols that underpin the IoT, and you'll be better equipped to understand the technical discussions around this exciting field. So, let's embark on this journey to unravel the technological underpinnings of the Internet of Things!

Communication Technologies

As you delve deeper into the world of the Internet of Things (IoT), the first aspect you'll encounter is communication technologies. These are the conduits that facilitate the flow of information among devices, from sensors to the cloud, and everywhere in between. Let's explore these technologies to understand how they enable the connectivity at the heart of IoT.

Wired Communication Technologies

While wireless communication often steals the spotlight in discussions around IoT, it's essential to acknowledge the role of wired communication technologies. They remain relevant, particularly in situations where reliability, speed, and security are paramount.

- **Ethernet:** Ethernet, a technology you're likely familiar with from home or office networks, is a standard for wired data networks. It's used extensively in industrial IoT applications, where devices are stationary, and the speed and reliability of a wired connection are crucial. For example, in a manufacturing setting, machinery might be connected via Ethernet to ensure real-time, uninterrupted data transmission for monitoring and control purposes.

- **Serial Communication:** In many low-cost, low-power IoT devices like Arduino-based setups, serial communication methods, such as UART (Universal Asynchronous Receiver/Transmitter), SPI (Serial Peripheral Interface), and I2C (Inter-Integrated Circuit), are utilized. These protocols are ideal for short-range, device-to-device communication.

Wireless Communication Technologies

Wireless communication has been a driving force in the expansion of IoT, providing the flexibility and scalability required for connecting billions of devices worldwide.

- **Wi-Fi:** One of the most ubiquitous wireless technologies, Wi-Fi, provides high-speed internet connectivity over short to medium ranges. It's the technology behind your wirelessly connected devices at home, such as your smartphone, smart TV, or home automation devices like smart thermostats or security cameras.

- **Bluetooth and Bluetooth Low Energy (BLE):** Bluetooth is another widely used short-range wireless communication technology. You see it in action when you connect your wireless headphones to your phone or when a health wearable device syncs data to an app. BLE, a power-efficient variant of Bluetooth, is particularly useful for IoT devices running on battery power, such as fitness trackers or smartwatches.

- **Zigbee and Z-Wave:** These are wireless protocols specifically designed for low-power, low-data-rate applications in IoT, home automation, and device-to-device communication in a local area. Examples include smart light bulbs or smart locks that use Zigbee or Z-Wave to communicate with a central home automation hub.

- **Cellular Technologies:** When you think of cellular communication, your mind likely jumps to the technology that connects your mobile phone to the internet. The same underlying technology, including 3G, 4G, and now 5G, is used in IoT for devices requiring long-range, high-speed, and reliable connectivity. For instance, a fleet of delivery trucks might use cellular connectivity for real-time GPS tracking across a city or even across the country.

- **LPWAN Technologies:** Low-Power Wide-Area Networks (LPWAN) technologies, such as LoRaWAN, Sigfox, and NB-IoT, offer long-range communication at a low data rate while consuming very little power. These are used for IoT applications where devices need to send small amounts of data over long distances, like a soil moisture sensor in a rural farm sending data to a central server for analysis.

Remember, the choice of communication technology in an IoT setup depends on several factors, including the data rate, range, power consumption, cost, and the specific requirements of the IoT application. As you delve deeper into IoT, understanding these technologies will provide a foundation for grasping the complex interplay of devices, data, and connectivity in the IoT ecosystem.

IoT Network Protocols

After understanding the communication technologies, the next step in our journey is to explore the Internet of Things (IoT) network protocols. These protocols define the rules for data transmission over a network, including the initiation and termination of connections, the format of the data packets, and the way devices respond to incoming data or requests. Let's delve into these protocols, their characteristics, and their application in real-world IoT scenarios.

HTTP (Hypertext Transfer Protocol)

HTTP is a widely known application protocol used primarily for transferring data over the internet. In the world of IoT, HTTP is used for its simplicity and wide compatibility. When your smart thermostat sends temperature data to the cloud, or when you use a web browser to access the dashboard of your IoT device, HTTP is likely the protocol facilitating that communication.

MQTT (Message Queuing Telemetry Transport)

MQTT is a lightweight messaging protocol designed specifically for low-bandwidth, high-latency networks, making it an excellent fit for many IoT applications. It follows a publish-subscribe model, where devices (clients) publish messages to topics, and other devices subscribe to these topics to receive the messages.

For instance, in a smart home, an MQTT broker could manage communication between various devices. Your smart thermostat might publish temperature readings to a topic, and your heating system, subscribed to that topic, would receive these readings and adjust its settings accordingly.

CoAP (Constrained Application Protocol)

CoAP is a simple, HTTP-like protocol designed for constrained devices (devices with limited processing and memory resources) typically found in IoT environments. It's useful for machine-to-machine (M2M) applications such as smart energy and building automation.

Consider a network of battery-powered environmental sensors in a smart building. These sensors could use CoAP to send data to a central server, benefiting from CoAP's low overhead and built-in support for discovery of services and resources.

DDS (Data Distribution Service)

DDS is a fast, scalable, and real-time data-centric connectivity protocol for large-scale and high-performance IoT applications. It's used widely in industries like aerospace, defense, and autonomous vehicles, where the speed of data delivery is critical.

For example, in an autonomous car, multiple sensors and systems need to communicate with each other in real-time. DDS could be the protocol facilitating this high-speed, low-latency communication, ensuring the car responds to its surroundings quickly and safely.

AMQP (Advanced Message Queuing Protocol)

AMQP is a robust, flexible protocol designed for reliable message delivery between applications or organizations. In an IoT context, it's often used in enterprise solutions where guaranteed message delivery is crucial, such as industrial IoT applications.

Imagine a large manufacturing plant with numerous IoT devices monitoring machinery. The data from these devices could be sent via AMQP to ensure no information is lost during transmission, maintaining accurate real-time monitoring and analytics.

LoRaWAN (Long Range Wide Area Network)

LoRaWAN is a network protocol designed for long-range, low-power communication. It's built on top of the LoRa (Low Range) wireless communication technology and is often used in wide area IoT applications.

Consider a smart city application where IoT sensors monitor environmental conditions across the city. These sensors could use LoRaWAN to send their data over long distances, with low power consumption, to a central server for analysis and action.

These are just a few examples of the myriad network protocols used in IoT. Each protocol has its own strengths and weaknesses, and the choice of protocol depends on the specific needs of your IoT application, such as power consumption, range, bandwidth, latency, and reliability. As you continue your journey into IoT, understanding these protocols will give you a deeper insight into how IoT devices communicate and interact, forming complex, interconnected networks that drive our smart homes, cities, industries, and more.

Zigbee and Z-Wave

Both Zigbee and Z-Wave are wireless protocols designed specifically for home automation. They operate on a mesh network structure, which means each device in the network can relay data for other devices, extending the range and reliability of the network. For instance, if you have a smart home setup with smart lights, a smart lock, and a smart thermostat, they could all be communicating with each other and the central hub using Zigbee or Z-Wave. Both protocols focus on low-energy consumption and secure communication, key requirements in a smart home environment.

6LoWPAN (IPv6 over Low-Power Wireless Personal Area Networks)

6LoWPAN is a network protocol that enables the efficient use of IPv6 over low-power, low-bandwidth wireless networks, like those used by many IoT devices. The primary advantage of 6LoWPAN is that it allows for direct IP connectivity between small, battery-powered devices and the wider internet. For example, a small, battery-operated temperature sensor in a field could use

6LoWPAN to send its data directly over an IPv6 network, simplifying the overall network architecture.

Thread

Thread is an IPv6-based, low-power, secure and future-proof networking protocol designed for home automation. It's designed to have low power consumption, robustness, and security. For example, if you have a smart home setup with various appliances like a refrigerator, a washing machine, a smart TV, and a home security system, Thread could be the protocol that enables these devices to communicate with each other and the internet in a secure and reliable manner.

Remember, in IoT, there's rarely a one-size-fits-all solution when it comes to choosing a network protocol. The right protocol for your IoT application will depend on a variety of factors, including the specifics of the devices being used, the nature of the data being transmitted, the required range and data rate, the power constraints, and the necessary level of security. By gaining a solid understanding of these protocols, you'll be well-equipped to make informed decisions as you design, build, or work with IoT systems in your journey.

Data and Message Protocols

Having learned about the underlying network protocols that govern how IoT devices communicate with each other and with the internet, the next topic to explore is data and message protocols. These are the rules that determine the structure and format of the data that is transmitted between devices and systems in an IoT network. Understanding these protocols is essential for ensuring seamless and meaningful communication in an IoT ecosystem.

MQTT (Message Queuing Telemetry Transport)

As we touched on earlier, MQTT is a lightweight, publish-subscribe protocol that's excellent for many IoT applications, particularly those involving low-power, bandwidth-constrained devices. MQTT messages contain a topic, which identifies the content or purpose of the message, and a payload, which is the actual data being transmitted.

For example, in a smart farming application, an IoT device monitoring soil moisture might periodically send MQTT messages with the topic "Farm1/Field2/SoilMoisture" and a payload containing the actual moisture reading. Any other devices or systems subscribed to this topic would receive these readings and could act upon them, like an irrigation system automatically watering the field when the soil moisture drops below a certain level.

CoAP (Constrained Application Protocol)

CoAP, like MQTT, is a protocol designed for constrained IoT devices. However, unlike MQTT, which is message-oriented, CoAP is more akin to HTTP, being a request-response protocol. This means a client device sends a request to a server device, which then responds with the requested data.

As an example, consider a smart home setup where a CoAP server is running on a home automation hub, and various IoT devices, like lights and thermostats, are the clients. You could use a smartphone app to send a CoAP GET request to the hub for the current temperature reading from the thermostat. The hub would then send a CoAP request to the thermostat, receive the temperature data, and send it back to your smartphone in a CoAP response.

HTTP/2 and HTTP/3

While HTTP is familiar to many as the protocol that powers the web, its newer versions, HTTP/2 and HTTP/3, bring several improvements and features that can be beneficial for IoT applications. These include multiplexing, which allows multiple requests and responses to be sent simultaneously over a single connection, and header compression, which reduces the amount of data needed to be sent for each request or response.

For example, if you have an IoT application where a device needs to send multiple different types of data to a cloud server simultaneously, like temperature, humidity, and pressure readings, HTTP/2 or HTTP/3 could allow these data to be sent in parallel over a single connection, reducing the overall transmission time and network overhead.

AMQP (Advanced Message Queuing Protocol)

AMQP, as mentioned earlier, is a protocol that provides robust, flexible, and reliable message delivery between applications. It's designed with a focus on interoperability and supports a range of messaging patterns, including publish-subscribe and request-response.

For instance, in an industrial IoT setup, an AMQP network could be used to relay data from various sensors and machinery to a central monitoring system. The data could include readings like temperature, pressure, speed, etc., and the AMQP network would ensure that all these messages are reliably delivered, even in the event of network disruptions.

XMPP (Extensible Messaging and Presence Protocol)

XMPP is a communication protocol primarily used for instant messaging and presence detection. However, it's also been adopted for IoT applications due to its flexibility, extensibility, and wide support for various features, like multi-party messaging, publish-subscribe, and secure communication.

Imagine you have a network of IoT devices in a smart building, like light sensors, occupancy sensors, and smart lights. These devices could use XMPP to communicate, sending messages to each other to indicate their status or to command actions. For example, an occupancy sensor could send a message when it detects someone entering a room, triggering the smart lights to turn on.

DDS (Data Distribution Service)

We mentioned DDS earlier as a network protocol, but it's also a data-centric publish-subscribe messaging protocol. DDS allows real-time communication between devices, making it ideal for high-performance IoT applications where timely delivery of data is critical.

Consider an autonomous vehicle, which needs to process vast amounts of sensor data in real-time to make driving decisions. The vehicle's systems could use DDS to share this sensor data, ensuring that it's distributed quickly and efficiently to the parts of the system that need it.

Protobuf (Protocol Buffers) and JSON (JavaScript Object Notation)

While Protobuf and JSON are not communication protocols like the others we've discussed, they're worth mentioning as they are common formats for structuring the data that's sent between IoT devices.

Protobuf is a binary protocol developed by Google. It's small, fast, and efficient, making it suitable for IoT devices with limited resources. JSON, on the other hand, is a text-based format that's easy to read and write, and is widely supported across different programming languages.

For example, an IoT device might use JSON to send sensor readings to a cloud server, where the data could be easily processed and visualized. Or, for more resource-constrained devices or high-performance applications, Protobuf could be used to structure the data, reducing the size of the data packets and the time taken to encode and decode them.

Understanding these data and message protocols will allow you to be well-equipped to design and manage IoT systems, ensuring that your devices can communicate effectively and that their data can be understood and utilized. As with the network protocols, the choice of data and message protocol will depend on the specific needs of your IoT application, including factors like the capabilities of your devices, the nature and volume of the data being transmitted, the required speed and reliability of communication, and the need for interoperability with other systems or devices.

Device Management Protocols

Having covered the network and data protocols that enable IoT devices to communicate with each other and the wider world, let's now turn our attention to device management protocols. These protocols are vital for maintaining, updating, and monitoring IoT devices throughout their lifecycle. From provisioning new devices to troubleshooting issues, device management protocols ensure that your IoT network runs smoothly and efficiently.

Lightweight M2M (LwM2M)

Lightweight M2M, often referred to as LwM2M, is a device management protocol specifically designed for IoT and Machine-to-Machine (M2M) applications. It provides a range of services, including device provisioning, firmware updates, fault management, and more.

Suppose you are managing an IoT network for an agricultural company with hundreds of soil sensors scattered across various fields. With LwM2M, you can remotely monitor the status of each device, push firmware updates to the sensors, or troubleshoot any issues that arise, all from a centralized management console. This greatly simplifies the task of maintaining such a large and geographically dispersed network.

Open Mobile Alliance Device Management (OMA DM)

The OMA DM protocol is a device management protocol developed by the Open Mobile Alliance. It's used for managing mobile devices, but it's also applicable to IoT devices. OMA DM provides services like provisioning, firmware updates, and device diagnostics.

Let's say you're running a fleet management company with hundreds of trucks equipped with IoT devices for tracking location, speed, fuel consumption, etc. Using OMA DM, you can manage these devices remotely, ensuring they are functioning correctly, updating their software as needed, and diagnosing and resolving any issues that may arise, thereby minimizing downtime and ensuring your fleet operates as efficiently as possible.

TR-069 (CWMP)

TR-069, also known as the CPE WAN Management Protocol (CWMP), is a device management protocol primarily used by Internet Service Providers (ISPs) to manage devices on their network, like modems and routers. However, its use has expanded to include IoT devices as well. TR-069 allows for remote configuration, software/firmware upgrades, status and performance monitoring, and troubleshooting.

Imagine you're responsible for a network of smart meters in a city. With TR-069, you can monitor the status and performance of each meter, push firmware updates, or troubleshoot issues remotely. This can significantly streamline operations and ensure the reliable functioning of your smart meter network.

SNMP (Simple Network Management Protocol)

SNMP is a protocol designed for network management, and while it's not specific to IoT, it can be used for managing IoT devices. SNMP allows you to monitor and control devices on a network, and it's widely used for managing both small and large networks.

For example, if you're running a smart building with various IoT devices like HVAC systems, security cameras, and access control systems, you can use SNMP to monitor the status and performance of these devices. If a device starts behaving unusually, SNMP can send you an alert so you can investigate the issue. Similarly, you can use SNMP to configure the devices, adjust their settings, or even reboot them remotely if needed.

Bluetooth Low Energy (BLE) GATT (Generic Attribute Profile)

While you might be more familiar with Bluetooth for connecting your phone to your wireless headphones or car stereo, BLE's GATT is also used for managing and communicating with IoT devices. BLE GATT allows devices to expose their data and functionalities in a structured manner, making it easy to interact with the device and control its operations.

Consider a wearable fitness tracker as an example. The tracker might use BLE GATT to expose various data like heart rate, step count, and battery level, as well as functionalities like starting a workout or setting an alarm. Your smartphone can then connect to the tracker via Bluetooth, discover these data and functionalities using GATT, and interact with the device accordingly.

Zigbee Cluster Library (ZCL)

We discussed Zigbee earlier as a network protocol, but it's also worth mentioning its device management capabilities. The Zigbee Cluster Library (ZCL) defines a set of "clusters" that represent common device functionalities and attributes, like on/off control, temperature measurement, or light color control. This makes it easy to discover and control the functionalities of a Zigbee device.

For example, in a smart home setup with various Zigbee devices like lights, smart plugs, and a thermostat, each device would expose its functionalities and attributes through ZCL. A central home automation hub could then discover these functionalities and control the devices accordingly, allowing you to automate various aspects of your home.

Device management is a critical aspect of IoT, and the choice of device management protocol will depend on the specific needs of your IoT application. Factors to consider include the type and capabilities of your devices, the scale and complexity of your network, the need for remote management and updates, and the level of interoperability required with other systems or devices.

Interoperability in IoT

Interoperability refers to the ability of diverse systems, devices, or applications to work together, or 'interoperate'. In the context of IoT, interoperability is the ability for your different IoT devices, regardless of their manufacturer or the protocols they use, to communicate and share data effectively with each other and with the wider system. Achieving interoperability can be challenging, but it's crucial for realizing the full potential of IoT. In this section, we'll delve into the challenges of achieving interoperability, strategies to overcome these challenges, and the importance of standards in ensuring interoperability.

Challenges in Achieving Interoperability

Interoperability is easier said than done. You might have devices from different manufacturers that use different protocols for communication. One device might use MQTT as its messaging protocol while another uses CoAP. Similarly, one device might use Zigbee for networking while another uses Wi-Fi. The devices need to be able to understand each other's messages and work together harmoniously for your IoT system to function as intended.

For example, imagine you're building a smart home system where the smart thermostat, smart lighting, and smart security system all need to interact. If the thermostat detects no one is home, it should be able to communicate with the lighting and security system to ensure the lights are off and the security system is armed. However, if these devices are from different manufacturers and use different protocols, achieving this seamless interaction can be challenging.

Strategies for Interoperability

So, how can you overcome these challenges and achieve interoperability? Here are a few strategies:

- **Use of IoT Gateways:** IoT gateways can serve as a bridge between devices that use different protocols. They can translate messages from one protocol to another, enabling devices to communicate effectively. In the smart home example, an IoT gateway could translate messages between the thermostat, lighting, and security system, allowing them to interact seamlessly.

- **Standardization:** Adopting standardized protocols can greatly improve interoperability. Instead of each manufacturer developing their own proprietary protocols, they can use widely accepted standard protocols, making it easier for devices to communicate.

- **Integration Platforms:** IoT integration platforms, like those provided by AWS IoT or Microsoft Azure IoT, can help manage and translate data from different devices and protocols. These platforms often provide tools for device management, data processing, and analytics, making it easier to handle data from a wide range of devices.

The Role of Standards

Standards play a vital role in ensuring interoperability. They define the rules for how devices should communicate, making it easier for different devices to understand each other. There are several organizations, such as the Internet Engineering Task Force (IETF), the IEEE, and the Zigbee Alliance, that develop and maintain these standards.

For example, the IETF maintains standards for protocols like HTTP and CoAP, while the IEEE has standards for network technologies like Wi-Fi and Bluetooth. By adhering to these standards, device manufacturers can ensure their devices are capable of communicating with other devices that follow the same standards.

However, with the rapidly evolving nature of IoT, new standards are continually being developed, and it's important to stay updated on these changes. Furthermore, while standards can greatly improve interoperability, they're not a magic solution. Different standards may still be incompatible with each other, and not all devices will necessarily adhere to these standards. Therefore, while standards are a crucial tool for achieving interoperability, they should be used in conjunction with other strategies like IoT gateways and integration platforms.

In conclusion, achieving interoperability in IoT can be challenging but is crucial for building effective and efficient IoT systems. Understanding the challenges and strategies for interoperability, and the role of standards, will allow you to be well-equipped to design and manage an IoT system where diverse devices work together seamlessly.

The Need for Cross-Domain Interoperability

As IoT evolves, we're seeing the rise of complex, cross-domain IoT systems. These systems span multiple industries or domains – such as transportation, healthcare, and agriculture – and require diverse devices to interact. For example, in a smart city, traffic sensors, pollution monitors, and public transport systems all need to share data to optimize traffic flow and air quality.

Achieving cross-domain interoperability can be even more challenging due to the greater diversity of devices and protocols involved. However, the same strategies discussed earlier – using IoT gateways, adopting standards, and leveraging integration platforms – can help achieve cross-domain interoperability.

Vendor Neutrality

Another key aspect of interoperability is vendor neutrality. This means that your IoT system should not be tied to a specific manufacturer or vendor. Instead, it should be able to work with devices from any vendor, as long as they adhere to the necessary standards or protocols.

Vendor neutrality gives you the flexibility to choose devices based on your specific needs, rather than being restricted to a particular vendor's offerings. It also allows you to add or replace devices more easily as your needs change or as new technologies become available.

For instance, let's say you're developing a smart farming system, and you've deployed soil moisture sensors from Vendor A. Later, you find that Vendor B has released new sensors that are more accurate or energy-efficient. If your system is vendor-neutral, you can easily incorporate the new sensors, even though they're from a different vendor.

Interoperability Testing

Finally, it's important to validate the interoperability of your IoT system through testing. Interoperability testing involves checking whether your devices and systems can effectively exchange and correctly interpret data. It's often done as part of the device or system's certification process.

You can conduct interoperability testing in various ways, such as simulating the IoT environment, testing devices in a lab setting, or conducting field trials. These tests can help you identify and resolve interoperability issues before deploying your IoT system, ensuring that it will function as expected.

In conclusion, interoperability is a key factor that can significantly influence the success of your IoT project. It requires careful planning, smart use of technologies, adherence to standards, and rigorous testing. By ensuring interoperability, you can develop an IoT system where diverse devices work together seamlessly, unlocking the full potential of IoT.

Protocol Selection Criteria

Choosing the right protocol for your IoT system is a critical decision that can significantly impact the system's performance, reliability, and overall success. The choice of protocol will determine how your devices communicate, how data is transmitted and received, and how the devices are managed. As such, you need to carefully evaluate your needs and choose a protocol that best fits your specific situation.

In this section, we'll explore the key factors that should guide your protocol selection process. We'll also look at practical examples to illustrate how these factors can influence your decision.

Data Transmission Needs

The first factor to consider is your data transmission needs. Different protocols are designed to handle different types of data and different data transmission patterns. For example, if your IoT system involves devices that need to continuously stream large amounts of data, protocols such as HTTP or MQTT might be suitable due to their robust data handling capabilities.

On the other hand, if your devices only need to transmit small amounts of data intermittently, a lightweight protocol like CoAP might be a better fit. As a practical example, if you're setting up a weather monitoring system where sensors send small packets of data intermittently, CoAP could be an ideal choice.

Power and Bandwidth Constraints

The power and bandwidth constraints of your devices should also influence your protocol selection. Some protocols, like HTTP, require more power and bandwidth due to their complex data structures and verbose messaging style. If your devices are battery-powered or operate in areas with limited network coverage, these protocols might not be suitable.

Protocols like MQTT, which uses a publish/subscribe model, or CoAP, which is designed for constrained environments, are much more efficient in their power and bandwidth usage. For instance, if you're deploying IoT devices in remote areas with limited power sources, such as wildlife tracking devices in a national park, a power-efficient protocol like MQTT or CoAP would be a wise choice.

Reliability Requirements

The reliability of data transmission is another important consideration. If your IoT system involves critical operations where data loss could have serious consequences, you need a protocol that ensures reliable data delivery. For instance, MQTT ensures reliable data transmission with its three quality of service (QoS) levels, allowing you to choose the appropriate level based on your needs.

In contrast, if your system can tolerate occasional data loss, a protocol with less stringent reliability guarantees might be acceptable. For example, in a system monitoring soil moisture levels for a small garden, occasional data loss might not have significant consequences, and a protocol with less stringent reliability guarantees might suffice.

Security Requirements

The security requirements of your IoT system should also guide your protocol selection. If your system handles sensitive data or controls critical operations, strong security features are essential. Protocols like HTTPS and MQTT over TLS offer robust security features, including data encryption and authentication mechanisms.

On the other hand, if your system does not handle sensitive data and is not a likely target for attacks, a protocol with lighter security features might be acceptable. However, it's important to note that any IoT system can potentially be a target for attacks, and some level of security is always recommended.

Scalability Needs

Lastly, consider the scalability needs of your IoT system. If you expect your system to grow over time, either in terms of the number of devices or the amount of data, you need a protocol that can scale effectively. Protocols like MQTT, with its lightweight messaging and efficient use of network resources, can handle large numbers of devices and high volumes of data.

For instance, if you're designing a system for a smart city project that could potentially involve thousands of devices and large volumes of data, a scalable protocol like MQTT would be a suitable choice.

In conclusion, choosing the right protocol for your IoT system involves carefully considering your specific needs and constraints. The data transmission needs, power and bandwidth constraints, reliability requirements, security needs, and scalability of your system are all critical factors that should guide your decision.

Remember, there's no one-size-fits-all solution when it comes to choosing a protocol for your IoT system. The best protocol for your system depends on your specific situation. The protocols we've discussed in this section – HTTP, MQTT, and CoAP – are just a few examples of the many protocols available for IoT systems. Each protocol has its strengths and weaknesses, and the key is to choose the one that best meets your needs.

Also, it's worth noting that you don't necessarily have to choose a single protocol for your entire IoT system. In some cases, it might make sense to use different protocols for different parts of your system. For example, you might use MQTT for device-to-cloud communication due to its efficient use of network resources, and HTTP for cloud-to-application communication due to its wide support and robustness.

Selecting an appropriate protocol, guided by key factors, is vital to making informed decisions that set your IoT system up for success.

Emerging Protocols and Technologies

As the field of IoT evolves and grows, new protocols and technologies are constantly being developed to address the unique challenges and opportunities that IoT presents. Staying informed about these emerging trends can help you make more informed decisions about your IoT system and take advantage of the latest innovations. In this section, we'll explore some of the emerging protocols and technologies that are shaping the future of IoT.

LPWAN Protocols

One of the most notable trends in IoT is the rise of Low Power Wide Area Network (LPWAN) protocols. LPWAN protocols are designed to provide long-range communication with low power consumption, making them ideal for IoT devices that need to transmit small amounts of data over long distances.

Two of the most prominent LPWAN protocols are LoRaWAN and NB-IoT.

LoRaWAN (Long Range Wide Area Network) is a protocol designed for wireless battery-operated devices. It provides long-range communication with low power consumption, making it ideal for IoT devices that need to transmit small amounts of data over long distances. For example, if you're deploying IoT sensors in a large agricultural field to monitor soil conditions, LoRaWAN could be a great choice due to its long range and low power consumption.

NB-IoT (Narrowband IoT), on the other hand, is a cellular-based protocol that provides wide coverage, low power consumption, and high device density. It's particularly well-suited for applications that require wide-area coverage, indoor penetration, and low cost. For instance, if you're implementing a smart metering solution for a utility company, where devices need to send data over long distances and penetrate buildings, NB-IoT could be the ideal choice.

5G and IoT

The advent of 5G is another emerging trend that's set to have a significant impact on IoT. 5G is the fifth generation of cellular network technology, and it offers several improvements over its predecessors, including higher data rates, lower latency, and the ability to connect many more devices simultaneously.

For IoT, these improvements can enable new use cases that were not previously feasible. For example, with 5G's high data rates and low latency, you could implement real-time video analytics for a security system, where video feeds from IoT cameras are analyzed in real time to detect potential threats.

Blockchain and IoT

The integration of blockchain technology with IoT is another emerging trend that offers exciting possibilities. Blockchain is a distributed ledger technology that provides a secure and transparent way to record transactions. When applied to IoT, blockchain can offer a way to securely and transparently track the data produced by IoT devices.

For instance, in a supply chain tracking system, IoT sensors could record data about the condition and location of goods, and this data could be recorded on a blockchain. This would provide a secure and transparent record of the goods' journey, which could be valuable for verifying the authenticity of the goods and ensuring that they've been properly handled.

AI and IoT

Artificial Intelligence (AI) is also playing an increasingly important role in IoT. AI algorithms can analyze the large volumes of data produced by IoT devices and extract valuable insights. These insights can be used to optimize the operation of the IoT system, predict future trends, and even automate decision-making.

For example, in a smart factory, AI algorithms could analyze data from IoT sensors to optimize production processes, predict maintenance needs, and detect anomalies. This could improve the efficiency of the factory, reduce downtime, and improve product quality.

In conclusion, the world of IoT is constantly evolving, with new protocols and technologies emerging all the time. By staying informed about these trends and understanding how they can benefit your IoT system, you can stay ahead of the curve and make the most of the opportunities that IoT offers.

Edge Computing and IoT

Edge computing is an approach that involves processing data closer to its source, rather than transmitting it to a central cloud for processing. This can greatly reduce latency, increase privacy, and decrease bandwidth usage - all of which are crucial for certain IoT applications.

Consider a scenario where you're using IoT devices for real-time surveillance in a security system. The devices need to analyze video feeds and quickly identify potential threats. With edge computing, these devices can process the video feeds locally, reducing latency and enabling quicker responses to potential threats.

Quantum Computing and IoT

Although still in its nascent stages, quantum computing holds the potential to revolutionize IoT. Quantum computers could process and analyze the massive amounts of data generated by IoT devices much more quickly and accurately than classical computers. While this technology is not yet mature, it's worth keeping an eye on as it could significantly impact IoT data analysis in the future.

Interoperability Standards

As the IoT ecosystem continues to grow, the need for interoperability between different devices and systems is becoming increasingly important. Several organizations are working on creating interoperability standards to ensure that devices from different manufacturers can communicate and work together seamlessly.

For instance, the Open Connectivity Foundation (OCF) is working on creating a set of standards for device discovery, connectivity, and device-to-device communication. By adhering to these standards when designing your IoT system, you can ensure that your devices will be able to communicate with other devices and systems, regardless of the manufacturer.

In conclusion, staying up to date with emerging protocols and technologies is crucial in the rapidly evolving world of IoT. Understanding these trends and how they can be applied to your IoT system, will allow. you to make more informed decisions and take full advantage of the benefits that IoT has to offer. From LPWAN protocols and 5G to blockchain, AI, edge computing, quantum computing, and interoperability standards - the future of IoT is exciting and full of potential.

Impact of 5G and Future Network Technologies on IoT

The emergence of 5G and future network technologies promises to usher in a new era for IoT, opening up a wealth of opportunities for more complex and data-intensive applications. In this section, we'll explore the impact of these technologies on IoT and how they can transform your IoT systems.

5G and IoT

5G is the fifth generation of mobile network technology, and it offers several significant improvements over its predecessor, 4G. Some of the most notable benefits of 5G for IoT include:

- **Increased Data Speeds:** 5G networks offer data speeds that are up to 100 times faster than 4G networks. This means you can transmit larger amounts of data more quickly, making it possible to use IoT devices for more data-intensive applications. For example, you could use IoT cameras to stream high-definition video for real-time video analytics, which could be invaluable for applications like security monitoring or traffic management.

- **Lower Latency:** 5G networks also offer significantly lower latency than 4G networks. Latency refers to the delay between when data is sent and when it's received. Lower latency means that your IoT devices can communicate more quickly and efficiently. This is crucial for applications that require real-time responses, such as autonomous vehicles, which need to quickly process and react to information about their surroundings.

- **Greater Device Density:** 5G networks can support a much larger number of devices in a given area compared to 4G networks. This is particularly important for IoT, as it allows you to deploy more devices and create denser IoT networks. This could be particularly beneficial in urban areas, where you might want to deploy a large number of IoT sensors for a smart city application.

- **Improved Energy Efficiency:** 5G networks are designed to be more energy-efficient than 4G networks. This can help extend the battery life of your IoT devices, which is crucial for devices that are deployed in remote locations or difficult-to-reach areas.

Future Network Technologies and IoT

While 5G is currently the cutting-edge network technology, researchers and engineers are already looking ahead to future generations of network technology. Although these future technologies are still in the early stages of development, they have the potential to bring even more significant improvements for IoT.

- **6G Networks:** 6G, the sixth generation of mobile network technology, is expected to offer even higher data speeds and lower latency than 5G. It's also expected to introduce new features, such as advanced AI capabilities and more robust security measures. These improvements could enable new IoT applications that aren't currently feasible.

- **Satellite IoT:** Satellite technology is being increasingly used to provide connectivity for IoT devices in remote areas where traditional network coverage is sparse. Companies like SpaceX and Amazon are launching satellite constellations to provide global broadband coverage, which could make it possible to deploy IoT devices in virtually any location on Earth.

- **Edge Computing:** While not a network technology per se, edge computing is a crucial component of future IoT networks. Edge computing involves processing data closer to its source, reducing the need for data to be transmitted long distances for processing. This can reduce latency, save bandwidth, and improve privacy and security, making it an essential part of many future IoT applications.

In conclusion, 5G and future network technologies promise to bring significant improvements for IoT, enabling more data-intensive applications, real-time responses, and denser IoT networks. Understanding these technologies and their benefits will allow you to make more informed decisions about your IoT systems and take full advantage of the opportunities these technologies offer. Whether you're planning a small-scale IoT deployment or a large-scale IoT network, these technologies can help you realize your vision and achieve your goals.

Module 2 Exam Questions

1. What is the primary purpose of communication technologies in IoT?

a) Data storage

b) Data processing

c) Data transmission

d) Data visualization

2. Which of the following is not a type of IoT network protocol?

a) HTTP

b) MQTT

c) CoAP

d) CSS

3. What is the main advantage of using MQTT protocol in IoT?

a) It supports high data speeds

b) It is ideal for high power devices

c) It is ideal for limited network bandwidth and power

d) It allows for easy data visualization

4. Which of these is not a data and message protocol for IoT?

a) AMQP

b) MQTT

c) CoAP

d) HTML

5. SNMP is a protocol used primarily for what purpose in IoT?

a) Data transmission

b) Device management

c) Data encryption

d) Data visualization

6. Which of the following is not a key consideration for interoperability in IoT?

a) Device compatibility

b) Data security

c) Protocol compatibility

d) Internet speed

7. The primary factor for selecting an IoT protocol should be:

a) The cost of the protocol

b) The popularity of the protocol

c) The specific requirements of the IoT system

d) The complexity of the protocol

8. Which of the following is not an emerging protocol or technology in IoT?

a) Blockchain

b) Edge Computing

c) Quantum Computing

d) GPS

9. How does 5G impact IoT?

a) Increases data speeds

b) Lowers latency

c) Supports a greater device density

d) All of the above

10. Which of the following network technologies is not linked to the future of IoT?

a) 4G Networks

b) 6G Networks

c) Satellite IoT

d) Edge Computing

11. What is the advantage of using LoRaWAN in an IoT network?

a) High data rate

b) High energy consumption

c) Long-range, low power requirements

d) All of the above

12. Which one is not an IoT device management protocol?

a) SNMP

b) LWM2M

c) DDS

d) SMTP

13. What is a significant advantage of edge computing in IoT?

a) Increased latency

b) Increased bandwidth usage

c) Reduced data processing time

d) Increased power consumption

14. Which of the following is not expected to be a feature of 6G networks?

a) Lower data speeds than 5G

b) Advanced AI capabilities

c) More robust security measures

d) Lower latency than 5G

15. What does the term 'interoperability' in IoT mean?

a) Ability of devices to operate on internal power

b) Ability of devices to operate independently

c) Ability of different devices to work together seamlessly

d) Ability of devices to work without human intervention

16. Why is DDS considered a real-time protocol?

a) It can transmit data across large distances

b) It can support a high volume of data

c) It ensures timely and reliable delivery of data

d) It reduces the power consumption of devices

17. What advantage does the Zigbee protocol offer for IoT applications?

a) High speed data transfer

b) High power consumption

c) Low cost and low power consumption

d) High latency

18. What does the term 'latency' in network communication mean?

a) Speed of data transfer

b) Delay between when data is sent and received

c) Amount of data that can be sent at once

d) Time it takes to encrypt data

19. Which of the following is not a feature of 5G networks?

a) Increased data speeds

b) Higher latency than 4G

c) Greater device density

d) Improved energy efficiency

20. What is the primary benefit of using satellite IoT?

a) High data speeds

b) Connectivity in remote areas

c) Lower energy consumption

d) Lower cost of implementation

Answers:

1. c
2. d
3. c
4. d
5. b
6. d
7. c
8. d
9. d
10. a
11. c
12. d
13. c
14. a
15. c
16. c
17. c
18. b
19. b
20. b

Module 3. IoT Devices and Sensors

Introduction to IoT Devices and Sensors

Welcome to Module 3 on IoT Devices and Sensors. In this module, you will embark on a fascinating journey into the heart of the Internet of Things – its devices and sensors.

To understand IoT, it's imperative to understand the devices and sensors that make it possible. These are the building blocks, the very foundations that enable the Internet of Things to come alive. They are the eyes, ears, and fingers of the IoT world, constantly sensing, monitoring, and acting upon the environment around us.

The IoT devices and sensors are crucial components in the IoT ecosystem. These tiny but powerful entities can monitor various environmental parameters, track location, sense motion, detect changes, and so much more. They are connected to the internet, allowing them to transmit the data they collect, which then can be processed and analyzed to make intelligent decisions.

Take, for example, a smart thermostat installed in your home. It's an IoT device equipped with sensors that can detect the temperature in your house, and based on your preferences, it can adjust the temperature for optimal comfort. Similarly, a wearable fitness tracker on your wrist measures your physical activity, heart rate, and even the quality of your sleep, providing valuable insights into your health.

However, IoT devices and sensors are not just limited to consumer applications. They have widespread usage in various sectors, including manufacturing, agriculture, healthcare, logistics, and many more. For instance, in the agriculture sector, IoT devices equipped with soil moisture sensors can help optimize irrigation, leading to improved crop yield and water conservation.

In this module, we will dive deep into the world of IoT devices and sensors. We will explore various types of devices, from smart home devices and wearables to industrial IoT devices. We'll discuss the types of sensors used in these devices and how they contribute to the functionality of the IoT system. We will also look at the role of edge computing in IoT and the important considerations for power management in IoT devices.

By the end of this module, you will have a comprehensive understanding of IoT devices and sensors, their functionalities, and their profound impact on various aspects of our lives and businesses. So, get ready to delve into the fascinating world of IoT devices and sensors, where the digital and physical worlds converge seamlessly.

Smart Devices and Wearables

We will begin by exploring how smart devices and wearables have revolutionized our interactions with technology and the environment around us.

Firstly, let's define what we mean by 'smart devices' and 'wearables'. A smart device is any electronic gadget that connects, shares, and interacts with its user and other devices. This interaction often happens over a network, frequently the Internet, enabling a level of sophisticated functionality. Wearables, on the other hand, are smart devices designed to be worn on your body. These include smartwatches, fitness trackers, smart glasses, and even smart clothing.

Now, let's explore these categories more deeply.

Smart Devices

The most common smart devices you may be familiar with are smartphones. Your smartphone is much more than a communication device. It's a camera, a GPS, a personal assistant, and even your personal banker. It connects to the internet, downloads data, and interacts with other devices, making it a key player in the IoT ecosystem.

But the realm of smart devices extends far beyond smartphones. In your home, devices like smart thermostats, smart lights, and smart refrigerators are transforming everyday living. Your smart thermostat learns your temperature preferences over time and adjusts your home's temperature automatically. Your smart lights can be controlled remotely or set to turn on as you arrive home. Your smart refrigerator can monitor its contents, alert you when you're running low on essential items, and even order groceries online for you.

Wearables

When it comes to wearables, the variety is truly remarkable. The most popular wearables are smartwatches and fitness trackers. Your smartwatch can monitor your heart rate, track your physical activity, provide you with notifications, and even allow you to make payments. A fitness tracker can help monitor your sleep patterns, track your steps, calculate calories burned, and provide personalized insights into your health and fitness.

Then there are smart glasses, which can display notifications, give directions, and even record what you see. There are even smart clothing and accessories such as jackets that can control your music player, shoes that give directions, and jewelry that alerts you to notifications on your phone.

Practical Examples

Imagine this - you wake up in the morning to the sound of your smart alarm, which is set to wake you up during your lightest sleep phase. As you get out of bed, your smart lights automatically adjust to a brighter setting. You check your smartwatch, which has tracked your sleep quality, and it suggests a light breakfast based on your diet plan and the morning workout you've scheduled.

After your workout, tracked and guided by your wearable device, you step into the shower. Your smart shower adjusts the water temperature based on your preference. As you get ready for work, your smart coffee maker already has your coffee ready, made exactly how you like it. Your smart mirror displays news updates, weather forecast, and even suggests the best outfit for the day based on your schedule and the weather.

As you leave your house, your smart home system turns off the lights, adjusts the thermostat, and locks the doors. Your smart car is ready, pre-heated, or cooled depending on the season, and suggests the best route to work based on real-time traffic data.

This scenario may seem futuristic, but all of these technologies exist today and are part of the IoT ecosystem.

IoT Sensors and Actuators

Now, you'll delve into the core components of any IoT system. We'll focus on the parts that allow your devices to interact with the world: sensors that perceive the environment, and actuators that make changes based on those perceptions.

IoT Sensors

In the simplest terms, a sensor in an IoT device is the component that detects changes in the physical environment. This could be anything from temperature fluctuations, to light levels, to changes in motion or pressure. The sensor then converts this physical data into an electrical signal which can be processed and interpreted by the device.

Let's take a deeper dive into some common types of sensors you'll find in IoT devices.

- **Temperature Sensors:** These are found in a wide variety of IoT devices, including smart thermostats, refrigerators, and even some wearables. They measure the degree of heat or cold and help the device adjust its operation accordingly. For example, your smart thermostat uses a temperature sensor to determine when to turn your heating or cooling system on or off.

- **Motion Sensors:** These are often found in security systems and smart lights. They detect movement in their vicinity and trigger a response. For instance, your smart lights may turn on automatically when you walk into a room, thanks to motion sensors.

- **Humidity Sensors:** These are often paired with temperature sensors, especially in agricultural IoT devices, to monitor moisture levels in the environment. A smart farming system might use humidity sensors to determine the best time to activate an irrigation system.

- **Pressure Sensors:** Found in various industrial IoT devices, these sensors measure the force exerted by a liquid or gas. In a smart factory, pressure sensors might monitor hydraulic systems to ensure they're operating safely.

- **Proximity Sensors:** These sensors detect when an object is near the device and are often found in smartphones and cars. For instance, a proximity sensor in your smartphone helps it to turn off the screen when you hold it to your ear during a call, saving battery life.

IoT Actuators

While sensors gather data from their surroundings, actuators perform actions based on this data. They're the components that allow your IoT devices to interact physically with their environment. Here are some examples of actuators:

- **Motors:** These convert electrical energy into motion. A motor in your smart blinds, for instance, allows them to open or close based on the time of day or light level detected by a sensor.

- **Pumps:** These move fluids and are often used in smart irrigation systems. A pump might be triggered to water your plants when a moisture sensor indicates the soil is too dry.

- **Speakers:** These convert electrical signals into sound, making them a type of actuator. Your smart speaker uses its speaker component to play music or give you a verbal weather update.

Practical Examples

Let's look at a day in your life with IoT sensors and actuators. As you wake up, your smartwatch, which has been tracking your sleep via motion sensors, detects that you're now awake. Based on this, it sends a signal to your smart coffee maker (an actuator) to start brewing your morning cup of coffee.

In the bathroom, the humidity sensor in your smart mirror notices a rise in moisture levels due to the hot shower you're taking. The mirror defogging system (an actuator) is automatically activated, clearing the mist from the mirror surface.

On your way to work, your smartphone's proximity sensor detects that it's close to your ear when you receive a call, and it automatically turns the screen off to save battery. At the same time, the light sensor on your phone notices you're outdoors in bright sunlight and automatically adjusts the screen's brightness for optimal visibility, enhancing your user experience.

Once at work, you attend a meeting in a conference room equipped with an IoT occupancy sensor. This sensor detects the presence of people in the room and sends a signal to the room's lighting and air conditioning systems (actuators) to adjust for optimal comfort and energy efficiency.

On your lunch break, you head to the gym. Your fitness tracker's heart rate sensor monitors your pulse during the workout, providing real-time feedback that helps you optimize your exercise routine. At the same time, the motion sensors in the equipment trigger the machine to adjust to your weight and movement for a personalized workout.

After work, as you approach your home, a smart security camera with a built-in motion sensor detects your presence. It signals the smart lock (an actuator) to unlock the front door, the smart thermostat to adjust the temperature to your preference, and the smart lights to illuminate your entryway.

As you prepare for bed, your smart home system, using a combination of time and light sensors, signals the motorized blinds (actuators) to close, the lights to dim, and the thermostat to lower the temperature for a comfortable sleep environment. Your wearable device, equipped with a variety of sensors, continues to monitor your heart rate, body temperature, and movement as you sleep, ready to start another day in your connected life.

These examples offer a glimpse into the extensive roles that sensors and actuators play in IoT devices. These components make our devices 'smart', enabling them to perceive and interact with the environment in ways that enhance our comfort, convenience, safety, and productivity. As you continue to interact with IoT devices in your daily life, take a moment to appreciate the sophisticated interplay of sensors and actuators that makes it all possible.

Edge Computing in IoT

As you continue to explore the fascinating world of IoT, you'll encounter an essential aspect that increases the power and efficiency of these connected systems - edge computing.

What is Edge Computing?

In the context of IoT, edge computing refers to the practice of processing data near the edge of your network, where the data is generated, rather than relying solely on a centralized cloud-based system. This means the data from your IoT sensors and devices is processed locally, either on the device itself or on a local server.

The main aim of edge computing is to move computation away from data centers towards the edge of the network, reducing the latency and bandwidth use of data transfers. This shift addresses the significant delays that can occur when sending data over long distances, as well as the costs and potential security risks associated with data transmission.

The Role of Edge Computing in IoT

Edge computing plays a crucial role in many IoT applications. By enabling data to be processed and analyzed at the edge of the network, it allows for real-time decision-making and actions, which are critical in certain IoT use cases. In addition, edge computing can help reduce the volume of data that needs to be transferred, stored, and processed in the cloud, leading to cost savings and efficiency gains.

Practical Examples

To help you better understand edge computing in IoT, let's explore a few practical examples:

- **Smart Home:** Imagine your smart home security system. It continuously monitors video feeds from several security cameras. If all raw video data were sent directly to the cloud for processing, it would consume significant bandwidth and potentially introduce latency. But with edge computing, the video can be analyzed locally, and only relevant events (like detecting a person at your front door) trigger data to be sent to the cloud. This local processing allows for real-time alerts and reduces the amount of data that needs to be transmitted and stored in the cloud.

- **Autonomous Vehicles:** In self-driving cars, quick decision-making is crucial. These vehicles generate vast amounts of data from various sensors, cameras, and radars. Any delay in processing this data could have serious consequences. Therefore, much of the data is processed right on the vehicle (at the edge), allowing for real-time reactions to the surrounding environment.

- **Wearable Devices:** Your smartwatch or fitness tracker constantly gathers health and fitness data. It can process much of this data locally, providing you with immediate feedback on your heart rate during a workout, for instance. Only the processed data that's needed for long-term analysis or storage might be sent to the cloud, optimizing bandwidth and preserving battery life.

- **Industrial IoT:** In a smart factory, machines and sensors generate vast amounts of data. Edge computing enables immediate analysis of this data, allowing for real-time quality control, predictive maintenance, and optimization of production processes.

By exploring these examples, you can see how edge computing makes IoT systems more responsive, efficient, and secure. It's an exciting area of IoT that continues to evolve and improve, shaping the future of how our devices will interact with us and the world around us. As you navigate your personal or professional IoT journey, understanding edge computing will undoubtedly be a valuable part of your toolkit.

Power Considerations for IoT Devices

In this section, you will delve into one of the essential, yet often overlooked, aspects of IoT systems: power management. Energy efficiency is a fundamental factor in the design, operation, and effectiveness of IoT devices. Let's explore this in more detail.

Why Power Considerations Matter

Power is a critical concern for IoT devices, particularly those running on batteries or energy harvesting technologies. With the exponential growth of IoT devices, many of which are meant to be deployed and operate autonomously for extended periods, efficient power usage becomes crucial. Optimizing power consumption can extend battery life, reduce environmental impact, and lower operational costs.

Factors Affecting Power Consumption

Several factors affect the power consumption of your IoT devices:

- **Device Functionality:** The more functions your device performs, the more power it consumes. For instance, a simple temperature sensor will use less power than a smart security camera with continuous video streaming.

- **Communication:** Transmitting data, especially over long distances or at high frequencies, can be a major power drain. Devices that communicate frequently, or that use high-power communication technologies, will have higher power consumption.

- **Data Processing:** On-device processing can consume significant power, especially for complex tasks. Edge computing can help balance the load between local processing (which consumes device power) and cloud processing (which saves device power but uses network resources).

- **Hardware and Software Design:** Efficient hardware components and well-optimized software can significantly reduce a device's power consumption. This includes everything from the choice of microcontroller to the design of the device's firmware and software.

Practical Examples and Strategies

Now, let's look at some practical examples and strategies for managing power in IoT devices:

- **Sleep Modes:** Many IoT devices spend most of their time doing nothing. A smart thermostat, for instance, only needs to check the temperature and adjust the heating or cooling system occasionally. Between these tasks, it can go into a low-power sleep mode, waking up only when needed.

- **Efficient Communication:** Reducing the frequency of data transmissions can save power. For example, a soil moisture sensor in a smart farm could be programmed to transmit data only when the moisture level changes significantly, rather than at fixed intervals.

- **Edge Computing:** As discussed in the previous section, processing data on the device can save power by reducing data transmissions. However, complex processing tasks can also drain power. The balance between edge and cloud computing will depend on the specific needs of your device and application.

- **Energy Harvesting:** Some IoT devices can supplement their power supply by harvesting energy from their environment. For instance, outdoor sensors might use solar panels, while a wearable device could generate power from your body heat or movements.

- **Battery Technology:** The choice of battery technology can impact the device's lifespan and performance. For example, a lithium-ion battery might provide high power and rechargeability for a smart home device, while a low-cost, long-life zinc-air battery could be the best choice for a simple environmental sensor.

By exploring these power considerations, you'll gain a more holistic understanding of IoT device design and operation. Whether you're a developer designing the next revolutionary IoT device or a user looking to optimize your smart home system, understanding power considerations is an invaluable part of your IoT journey.

Energy Harvesting Technologies in IoT Devices

At this stage of your IoT journey, you will be introduced to an innovative field that is revolutionizing how we power IoT devices - energy harvesting.

What is Energy Harvesting?

Energy harvesting, also known as power harvesting or energy scavenging, involves capturing small amounts of energy from the device's environment and converting it into usable electrical power. This technology offers an exciting solution for powering IoT devices, especially those deployed in remote or hard-to-reach locations where regular battery replacement is challenging.

Types of Energy Harvesting

Several different energy harvesting technologies are available, each capturing a different type of ambient energy. The choice of technology will depend on the specific environment and requirements of your IoT device. Let's explore some of these technologies:

- **Solar Energy Harvesting:** As you might guess, this involves capturing energy from sunlight or ambient light. Solar cells or photovoltaic cells convert light energy into electrical energy. Solar energy harvesting is popular for outdoor IoT devices or those located in well-lit indoor areas.

- **Thermal Energy Harvesting:** This method captures energy from heat, often in the form of temperature differences between two objects or areas. Thermoelectric generators, for example, can convert this heat difference into electricity. This method is useful for IoT devices located near heat sources or in environments with fluctuating temperatures.

- **Vibration Energy Harvesting:** Vibrational or kinetic energy, such as that from motion or mechanical vibrations, can be converted into electrical energy. This is often done using piezoelectric materials that generate a voltage when they're subjected to mechanical stress. This method can power IoT devices in industrial settings with lots of machinery or in wearable tech that captures energy from human movement.

- **RF Energy Harvesting:** Radio Frequency (RF) energy harvesting captures energy from ambient electromagnetic radiation, typically from nearby wireless communication systems. While the power generated by this method is often quite small, it can be enough for low-power IoT devices, like certain types of sensors.

Practical Examples

To help you better understand how energy harvesting works in real-world IoT applications, let's look at a few examples:

- **Solar-Powered Outdoor Sensors:** An outdoor environmental monitoring sensor, such as a weather station, could use a small solar panel to capture energy from the sun. This could power the sensor during the day, while a rechargeable battery could provide power during the night or on cloudy days.

- **Thermoelectric Generator in Industrial IoT:** In an industrial setting, a thermoelectric generator could be attached to a hot pipe or machine. The temperature difference between the hot surface and the cooler ambient air could be used to generate electricity, powering an IoT sensor that monitors machine performance or environmental conditions.

- **Vibration-Powered Wearable Device:** A wearable fitness tracker could incorporate a small piezoelectric generator that captures energy from your movements as you walk or run, helping to extend the device's battery life.
- **RF-Powered Smart Shelf:** In a retail setting, a smart shelf equipped with RFID tags could use RF energy harvesting to power the tags. The energy could be harvested from the signals sent by RFID readers, reducing the need for batteries.

Understanding the possibilities offered by energy harvesting technologies will allow you to start to see how they open up new opportunities for IoT deployment. As these technologies continue to evolve and improve, we can expect to see more IoT devices taking advantage of energy harvesting, leading to more sustainable and efficient IoT systems.

IoT Device Maintenance and Lifecycle Management

Here, you will learn about the important considerations in maintaining IoT devices, managing their lifecycle, and planning for end-of-life scenarios.

What is IoT Device Maintenance and Lifecycle Management?

Just like any piece of technology, your IoT device will need regular maintenance to ensure it functions optimally. This may include software updates, hardware checks, and troubleshooting any issues that arise. Lifecycle management, on the other hand, is about understanding and managing the phases your IoT device will go through from creation to disposal.

Phases of IoT Device Lifecycle

Typically, an IoT device goes through several phases in its lifecycle:

- **Design and Manufacture:** This is where your IoT device is born. The design should consider the target environment, user requirements, and integration with other devices or systems. The manufacturing process should then realize this design in a robust and reliable device.

- **Deployment:** In this phase, your IoT device is installed and starts its operational life. This might involve configuring the device, connecting it to the network, and testing to ensure it functions as expected.

- **Operation and Maintenance:** This is the longest phase of the lifecycle, where the IoT device performs its intended function. Regular maintenance will ensure the device continues to operate reliably. This might include tasks like checking battery levels, updating firmware, or calibrating sensors.

- **End-of-Life:** Eventually, your IoT device will reach the end of its useful life. This might be because the device is no longer needed, it's become obsolete, or it's broken and can't be repaired. In this phase, you'll need to consider how to dispose of or recycle the device in a responsible way.

Importance of Firmware Updates

One crucial aspect of IoT device maintenance is firmware updates. Firmware is the low-level software that controls your IoT device's hardware. Manufacturers often release firmware updates to fix bugs, patch security vulnerabilities, or add new features. Keeping your device's firmware up-to-date is essential for maintaining its functionality and security.

Device Longevity and End-of-Life Considerations

When considering the lifecycle of your IoT device, it's important to think about device longevity and end-of-life considerations. For example, how long do you expect your device to last? What will you do when the device is no longer needed or no longer works?

Some IoT devices might be designed to last for many years, especially those used in industrial settings or built into infrastructure. These devices might require robust materials and designs to withstand harsh conditions and long-term use.

On the other hand, some IoT devices might have a shorter expected lifespan. Wearable devices, for example, might be replaced every few years as technology improves and users upgrade to newer models.

When your IoT device reaches the end of its life, it's important to dispose of it responsibly. This might involve recycling the device, repurposing its parts, or returning it to the manufacturer for disposal.

Practical Example

Let's consider a practical example: a smart thermostat in your home.

- **Design and Manufacture:** The smart thermostat is designed to be easy to use, energy-efficient, and compatible with your home's heating and cooling system. It's manufactured with high-quality components to ensure reliable operation.

- **Deployment:** You install the smart thermostat in your home, connect it to your Wi-Fi network, and configure it using a mobile app. You might need to adjust some settings to suit your specific needs.

- **Operation and Maintenance:** The smart thermostat automatically adjusts your home's temperature based on your preferences, the time of day, or the outside temperature. You periodically check the device for any software updates and install them to ensure it's running the latest and most secure firmware. If you encounter any issues, such as the device failing to connect to the network or not responding correctly to temperature changes, you perform troubleshooting steps to identify and resolve the problem.

- **End-of-Life:** After several years, you decide to upgrade to a newer model of smart thermostat. You remove the old device and consider how to dispose of it. You could recycle it at a local electronics recycling facility, or you might check if the manufacturer offers a take-back program.

Understanding the maintenance needs and lifecycle of your IoT devices helps you plan for their long-term use and eventual replacement. This awareness can help you optimize the value you get from your IoT devices while minimizing potential disruptions and waste.

Proactive Maintenance and Predictive Analytics

With the power of IoT and data analytics, you can also move towards more proactive maintenance strategies. IoT devices can often provide data on their own performance, which you can analyze to predict potential failures before they occur.

For instance, an IoT-enabled air conditioning unit could monitor its own energy use, vibration levels, or other indicators of performance. By analyzing this data, you might notice patterns that indicate a potential failure, such as an increase in energy use or changes in vibration patterns. You can then perform maintenance or replace parts before the unit fails, avoiding unexpected downtime.

Device Management Platforms

To support you in managing and maintaining your IoT devices, various IoT device management platforms exist. These platforms can help you monitor the status of your devices, send firmware updates, manage network connections, troubleshoot issues, and more. They can be particularly helpful if you're managing a large number of IoT devices.

In conclusion, maintaining your IoT devices and managing their lifecycle is a critical part of your IoT journey. Understanding these processes and planning ahead will allow you to ensure your IoT devices deliver valuable insights and performance over the long term.

Case Studies of IoT Devices and Sensors

Finally, here, we will delve into some real-world examples of how IoT devices and sensors are put to work, highlighting the diverse applications and benefits they offer. This will provide you with concrete examples that will help you understand the concepts we've discussed so far in a more practical manner.

Case Study 1: Smart Agriculture - IoT in Farming

Imagine you're a farmer managing large tracts of land. One of the biggest challenges you face is knowing the right time to water your crops, how much fertilizer to use, and when to harvest for maximum yield. Enter IoT devices and sensors.

By using soil moisture sensors, you can accurately measure the water content in your soil. These readings, sent to a cloud-based platform in real-time, allow you to make data-driven decisions on when and how much to irrigate, reducing water waste and ensuring your crops get just the right amount they need.

In addition, by using other types of sensors such as weather stations, you can gather data on environmental conditions like temperature, humidity, and rainfall. This information can help you plan your farming activities more effectively.

Even more, IoT devices can monitor the health of your crops. For example, spectral analysis sensors can identify diseases or nutritional deficiencies in plants by detecting changes in how light is reflected off their leaves. By catching and addressing these issues early, you can improve your crop yield and quality.

Case Study 2: Predictive Maintenance in Manufacturing

Now imagine you're responsible for a manufacturing plant where any machine downtime can lead to significant financial losses.

Traditional maintenance practices involve routine checks that may not accurately predict failures.

With IoT sensors, you can shift to a predictive maintenance model. Sensors placed on your machinery can monitor various parameters such as temperature, vibration, and sound. This data is continuously analyzed, and using machine learning algorithms, it can predict possible failures before they occur.

For example, an increase in vibration levels or a sudden spike in temperature could indicate an impending failure in a machine part. By receiving this predictive data, you can schedule maintenance proactively, reducing downtime and saving costs.

Case Study 3: Smart Home Automation

In a different scenario, imagine you're a homeowner wanting to increase the comfort, convenience, and security of your home. IoT devices offer a plethora of options.

Smart thermostats can learn your schedule and preferences, adjusting your home's temperature automatically to save energy and improve comfort.

Connected security systems, which can include devices like smart locks, video doorbells, and motion sensors, can provide real-time alerts to your smartphone about any unusual activity, offering peace of mind.

Moreover, smart lighting systems can adapt to your routines and even change the ambiance of your rooms based on the time of day or your mood. Voice-activated assistants can control these systems and provide hands-free control of many elements of your home.

Case Study 4: Wearable Health Devices

Lastly, imagine you're a health-conscious individual who wants to keep track of your fitness levels and overall health. IoT wearable devices like smartwatches and fitness trackers make this easy.

These devices can track your heart rate, sleep patterns, activity levels, and more. You can use this data to understand your health trends, set fitness goals, and monitor your progress. Some wearables can even detect health anomalies like irregular heart rhythms and alert you or your healthcare provider.

Understanding these case studies can give you insights into the practical applications of IoT devices and sensors across various industries and aspects of life. As you delve deeper into the world of IoT, these examples can serve as inspiration and guide you in implementing IoT solutions in your own projects or industry.

Module 3 Exam Questions

1. What is the purpose of IoT device maintenance and lifecycle management?

a) To ensure the device functions optimally

b) To plan for the device's end of life

c) To fix bugs and patch security vulnerabilities

d) All of the above

2. Which of the following is not a typical phase of an IoT device lifecycle?

a) Design and Manufacture

b) Deployment

c) Operation and Maintenance

d) Marketing and Sales

3. Firmware updates are essential for:

a) Patching security vulnerabilities

b) Fixing bugs

c) Adding new features

d) All of the above

4. Which of the following is not a part of device longevity and end-of-life considerations?

a) How long the device is expected to last

b) How the device will be disposed of

c) The price of the device

d) What will be done when the device no longer works

5. In the context of IoT, what does edge computing enable?

a) Data processing on the device itself

b) Centralized processing of data

c) Slower response times

d) Increased network traffic

6. Energy harvesting technologies for IoT devices are used to:

a) Decrease device size

b) Increase device lifespan

c) Enable devices to self-generate power

d) Increase the cost of the device

7. Predictive maintenance in manufacturing involves:

a) Routine checks of machinery

b) Waiting for a machine to fail before repairing

c) Using sensors to monitor machine parameters and predict possible failures

d) Ignoring machine maintenance until absolutely necessary

8. In smart agriculture, soil moisture sensors are used to:

a) Detect diseases in plants

b) Determine when and how much to irrigate

c) Monitor crop yield

d) Test the quality of the soil

9. Which of the following is not a parameter monitored by IoT sensors in predictive maintenance?

a) Temperature

b) Sound

c) Vibration

d) Color of the machine

10. A smart thermostat in a home automation system can:

a) Learn your schedule and preferences

b) Adjust your home's temperature automatically

c) Save energy

d) All of the above

11. IoT wearable devices like smartwatches and fitness trackers can track:

a) Heart rate

b) Sleep patterns

c) Activity levels

d) All of the above

12. The use of spectral analysis sensors in smart farming helps to:

a) Identify diseases or nutritional deficiencies in plants

b) Predict the weather

c) Control pests

d) Increase soil fertility

13. In the context of IoT, what does LPWAN stand for?

a) Low-Power Wireless Area Network

b) Low-Power Wide Area Network

c) Large-Power Wide Area Network

d) Low-Power Wider Area Network

14. What is the purpose of a smart lock in a smart home?

a) To monitor temperature

b) To provide real-time security alerts

c) To control lighting

d) To control heating

15. Which of the following is not an example of a smart device or wearable?

a) Fitness tracker

b) Smart thermostat

c) Smart fridge

d) Traditional wristwatch

16. Why are IoT device management platforms important?

a) They monitor the status of devices

b) They help send firmware updates

c) They manage network connections

d) All of the above

17. Which of the following is not a benefit of edge computing in IoT?

a) Decreased latency

b) Reduced network traffic

c) Centralized data processing

d) Enhanced privacy and security

18. Which type of sensor would be most useful for a farmer wanting to optimize irrigation?

a) Vibration sensor

b) Soil moisture sensor

c) Light sensor

d) Temperature sensor

19. Which IoT device can detect irregular heart rhythms?

a) Fitness tracker

b) Smart fridge

c) Smart lock

d) Smart thermostat

20. Which of the following is a benefit of using IoT devices in manufacturing plants?

a) Improved product quality

b) Enhanced worker safety

c) Predictive maintenance

d) All of the above

Answers:

1. d
2. d
3. d
4. c
5. a
6. c
7. c
8. b
9. d
10. d
11. d
12. a
13. b
14. b
15. d
16. d
17. c
18. b
19. a
20. d

Module 4. IoT Data and Analytics

Introduction to IoT Data and Analytics

Welcome to Module 4 on IoT Data and Analytics. In this module, you will learn about the critical role that data plays in the Internet of Things ecosystem and how it fuels the decision-making process.

The Internet of Things (IoT) is a vast network of interconnected devices, each producing a wealth of data that, when harnessed effectively, can provide valuable insights and enable intelligent decision-making. But with this vast amount of data comes a set of unique challenges and opportunities. How do you collect and store this data effectively? How can you process and analyze it to extract useful information? These are the questions we aim to answer in this module.

We'll start by exploring the fundamentals of IoT data collection and storage. This will involve understanding the different types of data generated by IoT devices, as well as the various data storage options available. It's not just about collecting data but also about ensuring it is stored securely and can be accessed efficiently when needed.

Next, we'll delve into the world of data processing and analysis in the IoT landscape. You'll learn about the role of big data technologies and how they can handle the vast amounts of data generated by IoT devices. We'll also discuss the integration of machine learning and artificial intelligence (AI) in IoT, which is becoming increasingly important as we move towards more complex and intelligent IoT systems.

To give you a sense of the practical implications, imagine a city-wide network of IoT-enabled sensors installed to monitor air quality. These sensors generate a tremendous amount of data every second. Collecting, storing, and making sense of this data is a significant challenge. But with the right data and analytics strategies, this data can be used to identify pollution hotspots, inform policy decisions, and even predict future air quality conditions. This is just one of many examples of how IoT data and analytics can drive smarter decision-making and create value.

Finally, you'll learn about data visualization and dashboarding, which are essential tools for making sense of IoT data. Visualization tools allow us to see patterns, trends, and insights that might otherwise be difficult to discern from raw data.

By the end of this module, you'll have a solid understanding of the critical role data and analytics play in the IoT, and the tools and techniques used to extract value from this data. So let's begin our journey into the world of IoT data and analytics.

Data Collection and Storage

The first step in harnessing the power of IoT data lies in efficient data collection and storage. As you navigate the world of IoT, understanding how to collect and store data effectively will be pivotal for your projects. This section will guide you through this process, providing a comprehensive overview of IoT data collection and storage, along with practical examples.

Data Collection

Data collection in IoT is the process of gathering information from various IoT devices and sensors. Each device in your IoT network can generate a wide range of data types - from simple binary data (such as an on/off state from a light switch) to complex data (like a continuous stream of temperature readings from a weather sensor), and even multimedia data (like images from an IoT security camera).

So, how do you go about collecting this data? First, it's crucial to understand your data collection requirements, which will be largely dictated by the nature of your IoT project. For example, if you're building a smart thermostat system, you might need to collect data such as current temperature, humidity level, and the desired temperature setting.

Once you have defined your data collection requirements, the next step is to select the appropriate sensors or devices for data collection. In our smart thermostat example, this might involve using temperature and humidity sensors.

After deploying your sensors, you'll need to configure them to send the collected data to a central location, typically a database or a cloud-based storage system. This data transmission could be performed in real-time, at regular intervals, or based on certain events or triggers.

Data Storage

Once you've collected your data, you'll need a place to store it. Your choice of data storage will largely depend on the volume of data you're dealing with, how fast the data is being generated, and how you plan to use the data in the future.

If you're working with a small number of devices and relatively low data volumes, a simple database system might suffice. This could be a traditional relational database management system (RDBMS) such as MySQL, PostgreSQL, or SQLite, which can handle structured data effectively.

However, IoT data can often be unstructured or semi-structured, and the volume of data can be immense. This is where NoSQL databases come into play. NoSQL databases like MongoDB, Cassandra, and CouchDB are designed to handle large volumes of unstructured data and are often used in IoT scenarios.

You might also consider cloud-based storage solutions, such as Amazon's AWS IoT Core or Google's Cloud IoT Core. These platforms not only provide storage but also offer integrated tools for data processing and analysis, making them a convenient option for many IoT projects.

Let's go back to our smart thermostat example. The data collected from your temperature and humidity sensors might be stored in a time-series database. This type of database is ideal for storing and retrieving sequences of values over time, which is a common requirement in IoT scenarios.

In conclusion, effective data collection and storage is foundational to any successful IoT project. Understanding your data requirements, selecting the appropriate sensors and devices, and choosing the right storage solution, will allow you to lay the groundwork for meaningful data analysis and decision-making in your IoT initiatives. As you progress through your IoT journey, always remember that the goal is not just to collect and store data, but to use it to derive actionable insights and create value.

Data Processing and Analysis

Once you've collected and stored your IoT data, the next step is to process and analyze it. This step is where the real value of your data begins to shine through. By processing and analyzing your data, you can transform raw information into actionable insights.

Data Processing

Data processing in the context of IoT involves preparing your data for analysis. This can include tasks like data cleaning, data transformation, and data integration.

You might start with data cleaning, which involves identifying and correcting errors in the data, such as missing values or inconsistencies. For example, if you're collecting temperature data from a network of sensors, you might encounter a faulty sensor that reports an unrealistic temperature value. Data cleaning would involve identifying these erroneous readings and dealing with them appropriately.

Next, you may need to transform your data. This could involve normalizing data to a common scale, aggregating data points, or converting data to a different format. In our temperature monitoring example, you might want to aggregate the temperature readings into hourly averages or convert temperatures from Fahrenheit to Celsius.

Finally, data integration involves combining data from different sources. If you're collecting data from multiple types of sensors or devices, you'll need to bring this data together in a unified view. For example, in addition to temperature data, if you're collecting humidity data, you might integrate these two datasets to analyze the correlation between temperature and humidity.

Data Analysis

Once your data has been processed, it's ready for analysis. Data analysis involves applying statistical techniques and algorithms to your data to discover patterns, trends, and insights.

If you're just starting with your analysis, you might begin with descriptive analysis, which involves understanding the basic features of your data, such as mean, median, mode, and standard deviation. In our example, this could involve computing the average daily temperature or identifying the day with the highest temperature.

Next, you might move on to exploratory data analysis, which involves visualizing your data and looking for patterns or anomalies. For instance, you might plot a time series chart of your temperature data to identify any unusual spikes or dips.

As you delve deeper into your data, you might apply inferential or predictive analysis techniques. These techniques can help you understand relationships between variables and make predictions about future data points. For example, you might build a regression model to predict future temperature trends based on historical data.

In the realm of IoT, the sheer volume and velocity of data often necessitate the use of advanced analytics techniques such as machine learning and AI, which we will cover in more detail in the next section.

As a final step, you'll want to present your findings in a clear and understandable format. This might involve creating visualizations or reports that highlight your key findings and insights.

By mastering data processing and analysis, you're well on your way to unlocking the full potential of your IoT data. Remember, the ultimate goal is not just to collect data, but to use it to drive decision-making and create value. So, take the time to understand your data, explore it thoroughly, and let it guide you towards meaningful insights.

Cloud and Edge Analytics

As you navigate the world of IoT data analytics, one crucial decision you'll need to make is where to process and analyze your data. There are two main paradigms to consider: cloud analytics and edge analytics. Each has its strengths and is suitable for different scenarios. Understanding these paradigms will help you make informed decisions about your IoT analytics strategy.

Cloud Analytics

Cloud analytics is a traditional approach where the data generated by IoT devices is transmitted to the cloud - remote servers hosted on the internet - for storage, processing, and analysis.

The cloud offers virtually unlimited processing power and storage capacity, allowing you to scale your IoT applications as needed. It's a perfect fit for complex analytics tasks that require substantial computational resources or large datasets, such as machine learning model training.

For example, consider a smart city project where IoT sensors are deployed across the city to monitor air quality. The sensors collect data and send it to the cloud. Once in the cloud, you can aggregate data from all sensors and perform sophisticated analytics to identify patterns, trends, and correlations. The results could then inform policy decisions or trigger alerts.

Edge Analytics

Edge analytics, on the other hand, involves processing data directly on the IoT device (the "edge" of the network) or a local gateway device. The aim is to reduce the volume of data that needs to be sent to the cloud and provide real-time or near-real-time analytics.

Edge analytics is particularly useful for IoT applications that require quick decision making, low latency, or operate in environments with limited connectivity. It's also beneficial when dealing with sensitive data that you might not want to transmit to the cloud due to privacy or security concerns.

For instance, imagine you have an autonomous vehicle equipped with various IoT sensors. The vehicle needs to process data from these sensors in real-time to navigate its environment safely. Delaying processing until data is sent to the cloud and returned could result in accidents. Instead, the vehicle uses edge analytics to process sensor data on-board, making instant decisions about speed, direction, and braking.

Balancing Cloud and Edge Analytics

While cloud and edge analytics each have their merits, they're not mutually exclusive. In many cases, you'll find the best results come from a balanced, hybrid approach.

For example, in a factory using IoT devices for predictive maintenance, you might use edge analytics to monitor machine health in real-time, detecting and addressing critical issues immediately. At the same time, you could send a subset of the data to the cloud, where you can perform more detailed analysis to identify trends, predict future maintenance needs, and optimize operations.

In deciding between cloud and edge analytics, consider factors such as:

- **Data volume:** If your IoT devices generate a large volume of data, it might be impractical or costly to send all of it to the cloud. Edge analytics can help filter and reduce the data to a manageable size.

- **Connectivity:** If your IoT devices operate in areas with limited or unreliable connectivity, edge analytics can ensure that they can still function and provide value even when disconnected.

- **Latency requirements:** If your IoT application requires real-time or near-real-time decision making, edge analytics can provide the necessary speed.
- **Security and privacy:** If your IoT data is sensitive, processing it at the edge can reduce the risk of exposure during transmission.
- **Processing requirements:** If your analytics tasks are complex and require substantial computational resources, the cloud's processing power can be advantageous.

Navigating the trade-offs between cloud and edge analytics isn't always easy, but it's a crucial part of any IoT data strategy. Understanding these paradigms and carefully considering your needs and constraints, will allow you to design a solution that delivers valuable insights efficiently and effectively.

Real-time Analytics

As you delve deeper into the world of IoT data analytics, one term you'll frequently encounter is "real-time analytics". This concept refers to the ability to process, analyze, and act upon data as soon as it is created or received. Given the often time-sensitive nature of IoT data, understanding and implementing real-time analytics can be a game-changer for your IoT projects.

Understanding Real-Time Analytics

In traditional analytics, data is collected over a period of time, stored, and then processed in batches. While this method can handle large volumes of data and provide valuable insights, it often results in a delay between data generation and insight extraction.

Real-time analytics, on the other hand, minimizes this delay. As soon as an IoT device generates data, that data is immediately processed and analyzed. This ability to provide instant insights can be particularly valuable for time-sensitive applications where delays could have significant consequences.

Use Cases of Real-Time Analytics

To help you understand how real-time analytics works, let's consider a practical scenario. Suppose you're monitoring a smart home system with multiple IoT devices, including security cameras, smart locks, and motion sensors.

If an intruder tries to break into the home, your security camera might capture their image, your motion sensors might detect unusual activity, and your smart lock might register a forced entry attempt. With real-time analytics, your system can instantly process this data, recognize the threat, and trigger an alarm or alert to the homeowner or security company. If the system only processed data in batches, there could be a dangerous delay before the threat was recognized.

Implementing Real-Time Analytics

Implementing real-time analytics in your IoT projects involves several key steps:

- **Data Ingestion:** Your IoT devices need to be able to send data to your analytics system as soon as it's generated. This might involve using communication protocols that support real-time data transmission, such as MQTT or WebSockets.

- **Stream Processing:** Unlike batch processing, which deals with data in large, infrequent chunks, stream processing handles data in small, continuous streams. There are various stream processing frameworks available, such as Apache Kafka, Apache Flink, or Amazon Kinesis, that can help you handle this aspect.

- **Real-time Analytics Tools:** To analyze your data in real-time, you'll need tools capable of handling streaming data. These might include real-time databases, real-time analytics platforms, or even machine learning models trained to make instant predictions on streaming data.

- **Actionable Insights:** The goal of real-time analytics isn't just to produce insights faster; it's to act on those insights faster. This might involve setting up automated actions or alerts that are triggered by certain data conditions.

Challenges and Considerations

While real-time analytics can be incredibly powerful, it's not without its challenges. Real-time systems must be highly reliable and available to avoid missing crucial data. They also need to handle high data velocities and volumes, which can require significant computational resources.

Moreover, real-time analytics doesn't replace the need for batch processing. Batch processing is still useful for tasks that aren't time-sensitive or that require processing large volumes of historical data, such as training machine learning models or generating daily reports.

In your journey into real-time analytics, carefully consider the needs of your project. Do you need to respond to events as they happen, or can you afford to wait? What kind of infrastructure will you need to support real-time data processing and analysis? By considering these questions and understanding the principles of real-time analytics, you'll be well-equipped to unlock the full potential of your IoT data.

Big Data and IoT

The intersection of Big Data and IoT is a fascinating space, where the massive volumes of data generated by IoT devices meet the power of Big Data technologies. Understanding this relationship and how you can leverage it will be instrumental in your IoT journey.

What is Big Data?

Before we delve into the details, let's briefly recap what Big Data is. Big Data refers to extremely large datasets that are too complex to be handled by traditional data-processing software. These datasets are characterized by their volume, velocity, and variety - often referred to as the "3 Vs" of Big Data.

- **Volume:** This refers to the sheer size of the data. IoT devices can generate enormous amounts of data. For example, an autonomous vehicle can generate up to 4 TB of data per day, and a single jet engine can generate up to 500 GB of data per flight.

- **Velocity:** This refers to the speed at which data is generated and processed. IoT devices can often generate data in real-time or near-real-time, which requires a high-speed data processing capability.

- **Variety:** This refers to the different types of data that can be generated. IoT data can come in many forms, including structured data (like numerical data from a temperature sensor), semi-structured data (like JSON data from a smart device), and unstructured data (like images or video from a surveillance camera).

Big Data Technologies in IoT

With these characteristics in mind, it becomes clear that traditional data processing tools and techniques might fall short when dealing with IoT data. This is where Big Data technologies come into play. These technologies are designed to handle the volume, velocity, and variety of Big Data, making them well-suited for IoT scenarios.

For example, consider a scenario where you're managing a network of smart home devices. Each device in your network is constantly generating data - your smart thermostat is reporting temperature readings, your security cameras are capturing video footage, your smart fridge is logging usage patterns, and so on. The volume of data can be enormous, it's being generated in real-time, and it's coming in a variety of forms. To manage this data effectively, you might use a combination of Big Data technologies such as Apache Kafka for real-time data streaming, Hadoop for distributed data storage and processing, and Spark for fast, in-memory data processing.

Leveraging Big Data in IoT

Beyond just managing your data, Big Data technologies can also help you extract valuable insights from your IoT data. Advanced analytics tools like machine learning algorithms can be applied to your data to identify patterns, make predictions, and automate decision-making. For example, you could apply machine learning to your smart home data to predict future energy usage patterns and automatically adjust your thermostat settings to optimize energy consumption.

Additionally, Big Data technologies can enable real-time analytics in IoT. This means you can analyze and act upon your data as it's being generated, allowing for real-time decision-making. For instance, if your smart home system detects an unusual pattern in your security camera footage, it could send you an alert immediately.

In conclusion, the marriage of Big Data and IoT opens up a world of possibilities. As you explore the world of IoT, remember that the key to unlocking the value of your data lies not just in collecting it, but in harnessing the power of Big Data to analyze it and draw actionable insights.

Machine Learning and AI in IoT

As you delve deeper into the world of IoT, you'll quickly realize that Machine Learning (ML) and Artificial Intelligence (AI) play a critical role in maximizing the potential of IoT systems. The vast amounts of data generated by IoT devices can be overwhelming, but with the power of ML and AI, you can turn this flood of data into a wellspring of actionable insights.

Machine Learning in IoT

Machine learning is a subset of AI that provides systems the ability to learn and improve from experience without being explicitly programmed. In the context of IoT, machine learning can be used to identify patterns, make predictions, and automate decision-making based on the data generated by IoT devices.

Consider a scenario where you're using IoT sensors to monitor soil conditions in a smart farming setup. The sensors collect data on variables like temperature, humidity, pH level, and nutrient content. This data, while valuable, is just raw information until you apply machine learning algorithms to it. You could use supervised learning techniques to create a model that predicts the optimal watering schedule based on the sensor data. Or, you could use unsupervised learning to cluster the data and identify patterns that indicate the early stages of a crop disease.

Artificial Intelligence in IoT

While machine learning is primarily focused on making predictions based on known data, AI goes a step further to simulate human intelligence processes. AI can be used in IoT to automate complex tasks, make informed decisions, and even interact with humans in a more natural, intuitive way.

For example, imagine you have a smart home system equipped with AI. The system uses data from various IoT devices around the house, like security cameras, motion sensors, and smart appliances, and applies AI algorithms to interpret this data. It could identify unusual patterns in the data that suggest a security breach and automatically alert you or the local authorities. It could even use natural language processing (NLP) to understand your spoken commands and control your smart devices accordingly.

Deep Learning and IoT

Deep learning, a subfield of machine learning, is also making significant strides in the IoT landscape. Deep learning models, especially Convolutional Neural Networks (CNNs), are particularly useful in interpreting visual data from IoT devices such as surveillance cameras or industrial inspection systems.

Consider a scenario where you're using IoT cameras for quality control in a manufacturing line. The cameras capture images of the manufactured parts, and a deep learning model could analyze these images to identify defects. The model could be trained to recognize what a 'good' and a 'bad' part looks like and automatically flag any defective parts for review.

Leveraging ML and AI in IoT

As you can see, ML and AI have the potential to supercharge your IoT systems. However, leveraging these technologies requires a good understanding of your data, the appropriate ML or AI techniques, and the computational resources to train and run your models.

It's also important to remember that ML and AI are not magic bullets. They are tools that can help you extract more value from your IoT data, but they need to be applied judiciously and responsibly. Always consider the ethical implications of your ML and AI applications, especially when dealing with sensitive or personal data.

In summary, ML and AI enrich your IoT journey. Integrating these technologies can convert raw data to actionable insights, automate tasks, and create more natural device interactions.

Data Governance and Management in IoT

As you delve into IoT data analytics, you'll quickly realize that simply having vast amounts of data isn't enough. How you manage and govern that data is equally, if not more, important. Data governance refers to the overall management of the quality, consistency, usability, security, and availability of data in an organization. In the context of IoT, data governance and management become particularly critical due to the scale, complexity, and sensitive nature of the data involved.

Understanding Data Governance and Management

Data governance involves a set of processes and policies that ensure data is properly managed throughout its lifecycle. It covers aspects like data quality, data integration, data privacy, data security, and regulatory compliance. Effective data governance ensures that your IoT data is trustworthy, consistent, and ready for meaningful analysis.

Data management, on the other hand, refers to the technical practices and architectures used to handle data. This includes data collection, storage, backup, recovery, archiving, and disposal. Good data management practices ensure that your data is securely and efficiently stored, easily accessible, and not retained longer than necessary.

Implementing Data Governance and Management in IoT

Implementing effective data governance and management in your IoT projects requires careful planning and execution. Here are some key steps you might take:

- **Establish clear data governance policies:** Define who owns the data, who can access it, and what they can do with it. These policies should be clearly communicated and strictly enforced.

- **Implement robust data management systems:** Use database management systems (DBMS) suited for IoT use cases. These systems should be capable of handling the high velocity, volume, and variety of IoT data.
- **Ensure data quality:** Regularly clean your data and check for errors or inconsistencies. This will help to ensure that your analyses and decisions are based on accurate information.
- **Secure your data:** Implement strong security measures to protect your data both at rest and in transit. This could include encryption, secure protocols, and secure access controls.
- **Comply with regulations:** Make sure you're aware of and compliant with any data protection regulations applicable to your IoT data. This might include GDPR in Europe, CCPA in California, or other regional or sector-specific regulations.

Practical Examples

To help you understand these concepts, consider an IoT deployment in a healthcare setting, where various IoT devices are used to monitor patient health. The data generated is sensitive and protected by strict regulations, such as HIPAA in the United States.

As part of your data governance policy, you'd need to clearly define who owns this patient data (typically, the patient themselves), who can access it (e.g., specific healthcare professionals), and what they can do with it (e.g., use it for treatment decisions but not for marketing purposes).

On the data management side, you'd need to store this data securely, ensuring it's encrypted to protect patient privacy. You'd also need to have procedures in place for backing up data, recovering it in case of loss, and securely disposing of it when it's no longer needed.

Finally, you'd need to have checks in place to ensure data quality. This could involve checking for inconsistencies or errors in the data, such as unrealistic heart rate values or missing data points.

By implementing effective data governance and management practices, you can ensure that your IoT data is trustworthy, secure, and ready for analysis. This not only enhances the value you can derive from the data but also helps to protect the rights of the individuals or entities the data pertains to. As with all aspects of IoT data analytics, effective data governance and management require a thoughtful, systematic approach, but the rewards can be well worth the effort.

Integration of IoT Insights into Business Processes

As you navigate through the world of IoT data and analytics, you'll soon discover that collecting and analyzing data is only half the battle. The real challenge, and the true value of IoT, lies in integrating the insights derived from this data into your business processes. This process, often referred to as "operationalizing" your data, is a critical step in transforming raw data into actionable intelligence that can drive decision-making and business growth.

Understanding the Integration Process

Integrating IoT insights into your business processes is about making the insights derived from your IoT data actionable. This means using these insights to inform decision-making, optimize operations, improve customer service, or achieve other business objectives.

The process usually involves a combination of automated systems and human decision-making. For instance, insights derived from IoT data might trigger automated actions, such as adjusting the temperature in a smart home system or scheduling maintenance for a machine showing signs of wear. These insights can also be presented to decision-makers in a digestible format, like a dashboard, to inform strategic decisions.

Implementing IoT Insights Integration

Integrating IoT insights into your business processes involves several key steps:

1. **Define your objectives:** Before you can integrate IoT insights into your processes, you need to understand what you hope to achieve. Are you trying to improve operational efficiency, enhance customer service, reduce costs, or achieve other specific goals?

2. **Identify relevant insights:** Not all data is equally valuable. Focus on the insights that are most relevant to your objectives. These might include patterns, trends, or anomalies that can help you make more informed decisions or trigger important actions.

3. **Operationalize your insights:** This involves setting up systems to act on your insights, either automatically or by informing decision-making. For instance, you might set up an IoT system to automatically adjust machine settings based on sensor data, or you might use a data dashboard to keep your team informed about key metrics.

4. **Evaluate and optimize:** Once you've integrated your IoT insights into your business processes, monitor the results to see if you're achieving your desired outcomes. This might involve tracking key performance indicators (KPIs) or using other evaluation methods. Use these evaluations to continuously improve your processes.

Practical Examples

In the dynamic landscape of IoT, data plays a pivotal role in decision-making and process optimization. As you continue to explore the world of IoT Data and Analytics, this section will delve into several practical examples of how these concepts are applied in a real-world context. From predictive maintenance to quality control, energy efficiency, and workplace safety, you will gain a more comprehensive understanding of how data from IoT sensors, coupled with real-time analytics, can transform operations and yield significant benefits. Get ready to uncover the powerful potential of IoT in various scenarios, and how you can leverage it to enhance your own operational processes.

- **Predictive Maintenance:** Suppose your manufacturing plant includes several high-cost machines that are critical to your operations. Unexpected failures could lead to costly downtime. In this scenario, the IoT sensors could monitor machine health data such as temperature, vibration, and pressure. With real-time analytics, you can predict when a machine is likely to fail and perform maintenance proactively. This way, you avoid unscheduled downtime and extend the machine's life.

- **Quality Control:** You're manufacturing a product that requires precise assembly. IoT sensors are observing the assembly line, tracking key parameters like alignment, temperature, and pressure during the assembly process. Real-time analytics on this data can flag any deviations from the norm, allowing for immediate corrective action. This proactive approach reduces the number of defective products and ensures consistent quality.

- **Energy Efficiency:** Your plant is large and consumes significant energy. IoT sensors monitor energy consumption across different areas of the plant and at different times. By analyzing this data, you can identify patterns and trends, such as equipment that's consuming more energy than it should, or times of day when energy use is unusually high. Based on these insights, you can take measures to enhance energy efficiency, such as replacing or repairing inefficient equipment, and adjusting operations to off-peak energy hours.

- **Workplace Safety:** In a plant, safety is paramount. IoT sensors can monitor environmental conditions like temperature, humidity, noise levels, and presence of gases. If the temperature in a certain area rises above a safe level, or if a harmful gas is detected, alerts are sent in real-time, allowing for immediate evacuation and mitigation steps.

These examples showcase the power of IoT and real-time analytics in a practical, real-world context. By leveraging these technologies, you can optimize operations, improve safety, and increase overall productivity in your manufacturing plant.

Data Visualization and Dashboarding

As you collect, process, and analyze your IoT data, the final step in your data journey is to visualize your findings and present them in a way that is easy to understand and interpret. This is where data visualization and dashboarding come into play. These tools allow you to transform your raw data and analytical results into visual representations that can help you better understand your data and make informed decisions.

Data Visualization in IoT

Data visualization is the process of transforming data into graphical representations such as charts, graphs, or maps. In the context of IoT, data visualization can be a powerful tool for exploring your data, identifying patterns and trends, and communicating your findings.

For example, if you're using IoT sensors to monitor environmental conditions in different locations, you might have data on temperature, humidity, air quality, and so on. Instead of looking at these data points in a spreadsheet or database, you could create a line chart to visualize how these variables change over time. Or, you could create a heat map to visualize the spatial distribution of these variables across different locations.

The key to effective data visualization is to choose the right type of chart or graph for your data and the message you want to convey. For instance, line charts are great for showing trends over time, bar charts are useful for comparing categories, scatter plots can reveal correlations between variables, and so on.

Dashboarding in IoT

While data visualizations help you understand your data, dashboards help you monitor your data. A dashboard is a tool that aggregates multiple data visualizations into a single interface, allowing you to track key metrics and indicators in real-time.

In the IoT context, dashboards are often used to monitor the status and performance of IoT devices and systems. For instance, if you're managing a network of IoT devices in a smart home, you could create a dashboard that shows the current status of each device, historical usage patterns, alerts for any issues, and so on.

Creating an effective IoT dashboard involves identifying the key metrics and indicators that you want to track, designing a layout that presents this information in a clear and intuitive way, and setting up real-time data feeds to keep your dashboard up to date.

There are many tools available for creating data visualizations and dashboards, ranging from programming libraries like Matplotlib and Seaborn in Python, to specialized software like Tableau, Power BI, and Grafana. The tool you choose will depend on your data, your technical skills, and your specific needs.

Leveraging Data Visualization and Dashboarding in IoT

The true power of data visualization and dashboarding in IoT comes from the ability to turn raw data into actionable insights. By visualizing your data, you can identify patterns and trends that might not be apparent in raw data. And by monitoring your data in a dashboard, you can stay on top of your IoT system's performance and quickly respond to any issues.

However, creating effective visualizations and dashboards is not just about technical skills. It also requires a good understanding of your data and your audience. Always keep in mind who will be using your visualizations or dashboards and what information they need. Make your visualizations simple, clear, and focused on the key insights you want to convey.

In conclusion, data visualization and dashboarding are essential skills in your IoT toolkit. By mastering these skills, you can make your IoT data more understandable, actionable, and impactful.

Module 4 Exam Questions

1. What does IoT stand for?

a) Internet of Things

b) Interoperability of Technology

c) Integration of Telecommunications

d) None of the above

2. Which of the following is an example of an IoT device?

a) Smart thermostat

b) Fitness tracker

c) Connected car

d) All of the above

3. Which of the following is a key component of IoT architecture?

a) Sensors/Devices

b) Network Communication

c) Data Processing

d) All of the above

4. What is a primary function of IoT sensors?

a) To analyze data

b) To collect data

c) To delete data

d) To transmit data

5. Which of the following is a key aspect of data governance in IoT?

a) Data security

b) Data quality

c) Regulatory compliance

d) All of the above

6. In the context of IoT, what does data management refer to?

a) The purchasing of data

b) The technical practices and architectures used to handle data

c) The marketing of data

d) None of the above

7. Which of the following is NOT a step in integrating IoT insights into business processes?

a) Defining your objectives

b) Ignoring relevant insights

c) Operationalizing your insights

d) Evaluating and optimizing

8. What does it mean to operationalize your IoT insights?

a) To ignore them

b) To share them on social media

c) To make them actionable by integrating them into business processes

d) None of the above

9. Which of the following is a common IoT communication technology?

a) Bluetooth

b) Wi-Fi

c) Zigbee

d) All of the above

10. What is edge computing in the context of IoT?

a) The practice of processing data near the edge of your network, where the data is generated

b) The practice of storing data on the edges of the devices

c) The practice of collecting data from the edges of the network

d) None of the above

11. What is the role of machine learning in IoT data analytics?

a) To automate data deletion

b) To automate data collection

c) To automate data-driven decisions

d) None of the above

12. What is the purpose of data visualization in IoT?

a) To create artistic representations of data

b) To present data in a graphical, easy-to-understand format

c) To hide data from unauthorized users

d) None of the above

13. What is one of the main challenges of IoT security?

a) Ensuring interoperability

b) Ensuring data privacy

c) Ensuring data quality

d) All of the above

14. What is a key benefit of integrating IoT insights into business processes?

a) To improve operational efficiency

b) To enhance customer service

c) To reduce costs

d) All of the above

15. In the context of IoT, what does LPWAN stand for?

a) Low-Power Wide-Area Network

b) Limited-Power Wide-Area Network

c) Low-Power Wireless Access Network

d) None of the above

16. What is the main goal of data cleaning in IoT data analytics?

a) To ensure that analyses and decisions are based on accurate and relevant information

b) To make the data look more appealing

c) To reduce the amount of data being stored

d) None of the above

17. Which of the following is a real-time analytics example in the IoT context?

a) Adjusting the temperature in a smart home based on real-time data

b) Using historical data to predict future trends

c) Conducting an annual review of machine performance data

d) None of the above

18. In the context of IoT, what is the primary benefit of edge analytics?

a) It allows for faster decision-making by processing data closer to where it's generated

b) It makes the IoT devices look more sophisticated

c) It ensures that all devices are on the edge of the network

d) None of the above

19. Which of the following is an example of IoT in the healthcare industry?

a) A smart thermostat adjusting temperature based on weather forecasts

b) A wearable device tracking a patient's heart rate

c) A connected car providing real-time traffic updates

d) None of the above

20. What is the role of data governance in IoT?

a) Ensuring that data is managed in a way that complies with regulations

b) Ensuring that data is stored indefinitely

c) Ensuring that data is available to everyone

d) None of the above

Answers:

1. a
2. d
3. d
4. b
5. d
6. b
7. b
8. c
9. d
10. a
11. c
12. b
13. b
14. d
15. a
16. a
17. a
18. a
19. b
20. a

Module 5. IoT Security and Privacy

Introduction to IoT Security and Privacy

Welcome to Module 5 on IoT Security and Privacy. In this module, you will delve into one of the most critical aspects of the Internet of Things (IoT) - its security and privacy. As we bring more and more devices online, our digital footprint expands, raising essential questions about how we protect our data and systems.

Security and privacy are not mere add-ons or afterthoughts when it comes to IoT. They are integral to the successful deployment and acceptance of IoT technologies. With the increasing interconnectivity of devices, the surface area for potential attacks also expands, making IoT systems a tempting target for malicious entities.

Throughout this module, you will understand why these issues are of paramount importance. We will explore the unique challenges IoT presents to security and privacy, from the proliferation of network access points to the difficulties in standardizing security protocols across an array of devices.

We'll discuss examples of significant security breaches and how they could have been prevented. Consider, for example, a smart home system. It's convenient to control your home temperature, lighting, and security systems remotely. But what if a hacker gains control? They could manipulate your home environment or, worse, disable your security systems to gain physical access to your home. These are real threats that have occurred, and we will examine them in detail.

Additionally, we'll shed light on best practices for securing IoT systems. You'll learn about various cryptographic methods, authentication protocols, and other tools that developers and users can employ to ensure their IoT systems are secure.

Furthermore, we'll delve into privacy concerns. With IoT devices often collecting vast amounts of data, sometimes of a very personal nature, ensuring privacy is respected and legal standards are met is crucial. For instance, a fitness tracker collects health data, and a smart speaker might have access to private conversations. How do we ensure this data is used responsibly and kept private?

By the end of this module, you will have a comprehensive understanding of IoT security and privacy, armed with the knowledge to critically evaluate IoT systems and contribute to the development of safer, more secure, and respectful IoT technology.

This journey is not just about acquiring knowledge but about understanding the responsibility that comes with creating and using IoT technologies. So, let's embark on this journey of unraveling the complex world of IoT security and privacy.

Security Challenges in IoT

As you venture deeper into the world of the Internet of Things (IoT), you'll quickly realize that the vast network of interconnected devices brings with it a unique set of security challenges. These challenges arise from the inherent nature of the IoT landscape, which is marked by diversity in device types, communication protocols, and applications.

Scalability of Security Measures

With IoT, you're not dealing with a few dozen devices; you're often managing hundreds, thousands, or even millions of devices. Each device represents a potential entry point for malicious entities, and providing robust security measures that can scale to this level is a significant challenge. Imagine having to individually configure security settings for each of a thousand devices in an industrial IoT setting. The scale of IoT deployments often makes traditional manual security methods impractical.

Device Diversity and Interoperability

The IoT ecosystem is characterized by an extensive range of devices, from tiny sensors in a smart home to large machines in a factory setting. These devices often run on different operating systems, use various communication protocols, and have differing computational capabilities. Ensuring that these diverse devices can securely interact and exchange data without exposing vulnerabilities is a complex task.

Limited Computational Resources

Many IoT devices are designed to be compact and energy-efficient, which often means they have limited processing power and memory. These constraints make it difficult to implement robust security measures, such as encryption, which typically require significant computational resources. Therefore, striking a balance between security and device performance becomes a major challenge.

Insecure Networks

IoT devices often communicate over wireless networks, which are inherently more vulnerable to interception and interference than wired networks. An attacker could, for example, interfere with the communication between your smart door lock and your smartphone, potentially gaining unauthorized access to your home.

Physical Security

Unlike traditional computing devices like servers or desktop computers, which are typically housed in secure locations, many IoT devices are deployed in the field and are physically accessible. For instance, a smart meter outside your house could be tampered with, potentially leading to false readings or an interruption in the power supply.

Lack of Updates and Patch Management

Many IoT devices are designed for long-term deployment without regular maintenance. This makes the delivery of regular software updates and patches, which are crucial for maintaining security, a significant challenge. Without these updates, your devices can become vulnerable to new threats over time.

Data Privacy

IoT devices often collect personal or sensitive data, such as health information from a wearable device or voice commands from a smart speaker. Ensuring this data remains private and is not misused or accessed without authorization is a critical challenge in IoT security.

Lack of Standardization

The IoT industry suffers from a lack of universal security standards. Manufacturers often have their own set of security protocols, which might not be compatible with other devices in the network. This lack of standardization not only makes secure device interaction more challenging but also makes it harder for you to ensure every device in your network adheres to a high security standard.

As you can see, the security landscape in IoT is complex, with multiple overlapping challenges. But don't worry - understanding these challenges is the first step towards mitigating them. In the following sections, we'll explore strategies and best practices that you can use to secure your IoT devices and networks.

IoT Security Best Practices

Security is paramount in the IoT ecosystem. By following best practices, you can help prevent potential breaches and enhance the integrity of your IoT devices and networks. Let's explore some of these practices in detail.

Device Authentication and Authorization

Always ensure that every device in your IoT network is authenticated and authorized. This means each device should have a unique identity that can be verified before it's allowed to communicate with other devices or systems. For example, when adding a new smart light bulb to your home network, it should first prove its identity (authentication) and then be given specific permissions (authorization) about what it can and cannot do.

Implement Strong Encryption

Data, whether at rest or in transit, should be encrypted. This means that even if someone intercepts your data, they won't be able to understand it without the decryption key. Consider your smart thermostat: it collects data about your home's temperature and your usage habits and sends this data to the cloud for processing. If this data is transmitted without encryption, an attacker could intercept and read the data. But with strong encryption, the intercepted data would be useless to the attacker.

Regular Software Updates and Patch Management

Make sure to regularly update your IoT devices with the latest firmware and software patches. These updates often contain security enhancements and fixes for known vulnerabilities. For instance, if your smart security camera manufacturer releases a software update, it's crucial to install this update as soon as possible. This way, you can protect your camera from newly discovered threats or exploits.

Secure Network Communication

Your IoT devices should communicate over secure networks. This could mean using Virtual Private Networks (VPNs) for remote access or ensuring your devices use secure communication protocols like HTTPS. For example, when you're accessing your smart home hub remotely, a VPN could encrypt your connection, keeping your commands safe from prying eyes.

Least Privilege Principle

The principle of least privilege (PoLP) is a computer security concept in which a user is given the minimum levels of access necessary to complete his or her job functions. This principle should also be applied to your IoT devices. For instance, your smart fridge probably needs internet access to update weather forecasts or sync with your online grocery list, but it doesn't need access to your personal files on your home network. By limiting its privileges, you can reduce the risk if it ever becomes compromised.

Physical Security

Physical security is as important as digital security. Ensure your IoT devices are protected from physical tampering or theft. For example, if you have a smart meter outside your home, it might be worth installing it in a locked box or an inaccessible area to prevent physical tampering.

Privacy by Design

When dealing with IoT devices that collect personal or sensitive data, it's important to follow the principle of "privacy by design". This means that privacy should be a consideration at every stage of your IoT implementation, from the initial design of the devices to the data processing and storage stages. If you're using a health tracker, for instance, make sure that the data it collects is adequately protected, anonymized where possible, and not shared with unauthorized parties.

Security Audits and Testing

Finally, conduct regular security audits and penetration testing to identify potential vulnerabilities in your IoT system. This involves testing your devices and network as if you were an attacker, trying to find and fix security weaknesses before someone else does. For instance, you might try to "break into" your own smart home system to see if there are any vulnerabilities that you've missed.

By following these best practices, you'll be well on your way to creating a more secure IoT environment. Remember, security is not a one-time event but an ongoing process. Always stay informed about the latest security trends and threats in the IoT space.

Authentication and Encryption

In the vast and interconnected world of the Internet of Things (IoT), two security measures are paramount: Authentication and Encryption. These two practices form the foundation of most IoT security strategies, safeguarding your devices and the data they generate.

Authentication

Authentication is the process of verifying the identity of a device, user, or system. It's akin to checking a person's identification card before allowing them to enter a secured building. In the context of IoT, authentication ensures that only authorized devices can join your network and communicate with other devices.

For example, consider you've just bought a new smart light bulb. When you attempt to connect it to your smart home network, the system should first authenticate the bulb, ensuring it's a valid and trustworthy device. This could be done through a pre-shared key, digital certificates, or other authentication mechanisms.

Authentication in IoT can take several forms:

- **Password or PIN authentication:** This is the simplest form of authentication, where you enter a password or PIN to verify your identity. However, with the sheer number of IoT devices you might have, managing individual passwords can be cumbersome and insecure if not handled properly.

- **Two-Factor Authentication (2FA):** This involves using two different methods to verify your identity, usually something you know (like a password) and something you have (like a smartphone to receive a one-time password). 2FA provides an additional layer of security for your IoT devices.

- **Certificate-based authentication:** This method involves using digital certificates to authenticate devices. Each device is issued a unique certificate, and this certificate is used to verify the device's identity. This method is highly secure and scalable, making it suitable for large IoT deployments.

- **Biometric authentication:** For certain IoT applications, like smart locks or personal wearable devices, biometrics like fingerprints or facial recognition can be used for authentication. This offers a high level of security but requires more sophisticated hardware.

Encryption

While authentication ensures that only authorized devices can access your network, encryption protects the data that these devices send and receive. Encryption is the process of scrambling data into an unreadable format, which can only be reverted to its original form (decrypted) using a specific key.

Consider your smart thermostat that sends temperature data to a cloud server for analysis. Without encryption, anyone who intercepts the data while it's in transit could read and understand it. But with encryption, the intercepted data would be gibberish without the decryption key.

There are two primary types of encryption used in IoT:

- **Symmetric encryption:** In symmetric encryption, the same key is used to both encrypt and decrypt the data. It's like having a lock and key - the same key locks and unlocks the door. Symmetric encryption is fast and efficient, making it ideal for resource-constrained IoT devices. An example of symmetric encryption is the Advanced Encryption Standard (AES).

- **Asymmetric encryption:** In asymmetric encryption, two keys are used: a public key to encrypt the data and a private key to decrypt it. It's like having a mailbox where anyone can drop a letter (encrypt) using the publicly accessible slot, but only you can open (decrypt) it with your private key. Asymmetric encryption is more secure than symmetric encryption, but it's also more resource-intensive. A commonly used form of asymmetric encryption is the RSA algorithm.

By combining robust authentication and strong encryption, you can significantly enhance the security of your IoT devices and networks, ensuring that only authorized devices can connect to your network and that the data they handle remains confidential and intact.

Data Privacy and Compliance

As you interact with an ever-growing number of IoT devices, from smart speakers to fitness trackers, these devices collect and process a significant amount of data. Some of this data can be quite personal or sensitive, such as your health data or your location. Thus, ensuring the privacy of this data and complying with relevant regulations is a critical aspect of IoT security.

Data Privacy

Data privacy refers to the practices and strategies put in place to ensure that the data collected, processed, and stored by your IoT devices is only used in ways that respect your privacy. This involves several key principles:

- **Data Minimization:** Your IoT devices should only collect the minimum amount of data necessary to perform their function. For example, a smart thermostat needs to know the current temperature and your preferred settings but doesn't need to know your name or email address.

- **Purpose Limitation:** The data collected by your devices should only be used for the purpose for which it was collected. If your fitness tracker collects heart rate data to monitor your exercise, that data should not be used for other purposes, such as targeted advertising, unless you've explicitly agreed to this.

- **Access Control:** Only authorized individuals or systems should have access to your data. For instance, the data from your smart home security system should only be accessible to you and not to the manufacturer or any third parties.

- **Data Protection:** Your data should be protected at all times, both while it's being transmitted (in transit) and when it's stored (at rest). This often involves encryption, as well as other security measures like firewalls and secure access controls.

Compliance

In addition to these privacy best practices, you must also consider compliance with relevant regulations. These regulations vary by region and sector, but they generally aim to protect consumers' privacy and set standards for data security. Here are a few key regulations to be aware of:

- **General Data Protection Regulation (GDPR):** If you're in the European Union or if your devices collect data from EU citizens, you need to comply with the GDPR. This regulation has strict rules about data consent, data rights (like the right to access your data or have it deleted), and data protection.

- **California Consumer Privacy Act (CCPA):** For those in California or dealing with data of California residents, the CCPA applies. Similar to the GDPR, the CCPA gives consumers more control over their personal data.

- **Health Insurance Portability and Accountability Act (HIPAA):** If your IoT devices deal with health data, such as a wearable health monitor or a smart hospital bed, you might need to comply with HIPAA. This U.S. regulation sets standards for the protection of health information.

- **Payment Card Industry Data Security Standard (PCI DSS):** If your devices handle credit card information, such as a smart vending machine or a point-of-sale system, PCI DSS compliance is necessary. This standard aims to secure credit card data against theft and fraud.

Understanding and navigating these regulations can be complex, but they're an essential part of IoT data privacy and security. Always ensure you're aware of the regulations that apply to your specific IoT devices and that you're taking the necessary steps to comply.

In conclusion, data privacy and compliance are not just about checking boxes or avoiding fines. They're about building trust with users and fostering a more secure and privacy-respecting IoT ecosystem. By putting data privacy and compliance at the forefront of your IoT strategy, you're showing respect for your data and the individuals it represents.

IoT Security Case Studies

Examining real-world case studies can be one of the most effective ways to understand the challenges and solutions in IoT security. It can give you insights into the types of attacks that have happened, how they were dealt with, and most importantly, how they could have been prevented. Let's dive into a few practical examples.

Case Study 1: The Mirai Botnet

The Mirai botnet was one of the most notorious IoT-based cyberattacks. In 2016, this malware targeted online consumer devices such as IP cameras and home routers, infecting them by using default credentials that users had failed to change. Once infected, these devices were used to launch distributed denial of service (DDoS) attacks, one of which took down a large portion of the internet, including Twitter, Netflix, and Reddit.

The Mirai botnet is a stark reminder of the importance of basic security practices, such as changing default usernames and passwords on your IoT devices. It's also a demonstration of the impact of a DDoS attack, where a network of infected devices can be used to overload servers and disrupt online services.

Case Study 2: Smart Home Hacks

There have been numerous instances where smart home devices have been hacked, often due to weak security measures. In one case, a family's smart home was hacked, with the attacker gaining access to the smart thermostat, cameras, and even the smart lock on the front door. The hacker was able to spy on the family through the cameras and alter the home's temperature by adjusting the thermostat.

This example emphasizes the importance of secure network communication and device authentication in your IoT ecosystem. Using a secure network, regularly updating device software, and enabling two-factor authentication where possible could have prevented unauthorized access.

Case Study 3: St. Jude's Cardiac Devices

In 2017, the U.S. Food and Drug Administration (FDA) confirmed that certain cardiac devices made by St. Jude Medical were vulnerable to cybersecurity attacks. The vulnerability could have allowed a hacker to deplete the battery or administer incorrect pacing or shocks, putting patients' lives at risk.

This case study underscores the critical importance of IoT security in healthcare, where the stakes are incredibly high. Security must be built into these devices from the ground up, with ongoing software updates to address potential vulnerabilities. Furthermore, it highlights the importance of regulatory compliance, such as HIPAA in the U.S., which sets standards for protecting health information.

Case Study 4: Jeep Cherokee Hack

In 2015, two security researchers demonstrated that they could remotely hack a Jeep Cherokee, taking over the car's engine, brakes, and even steering. The hackers exploited a vulnerability in the car's entertainment system to gain access to the car's more critical systems.

This case shows the potential risks associated with the Internet of Vehicles (IoV) and emphasizes the need for strong security measures at all levels of IoT systems. The principle of least privilege could have been applied here, where the entertainment system should not have been able to access the car's driving systems.

By reviewing these case studies, you can gain a better understanding of the potential security risks associated with IoT devices and the best practices to prevent such attacks. Always remember, the key to IoT security lies in a proactive approach, anticipating potential vulnerabilities, and implementing strong security measures.

Security by Design

Incorporating 'Security by Design' in IoT means ensuring security from the initial stages of designing your systems and devices, rather than as an afterthought.

In this section, you will understand the importance of Security by Design and learn how to implement it in your IoT projects. You will also get acquainted with key principles that underpin this concept, each illustrated with practical examples.

Firstly, let's understand what 'Security by Design' means. It's a proactive approach to security, implying that you consider and implement security measures right from the design phase of your IoT product or system. This shift in perspective makes security an integral part of the IoT device or system's lifecycle, instead of being a reactive measure or an added layer after the system is already in place.

Principle 1: Least Privilege

The principle of least privilege means that a user or a program should have the minimum levels of access—or privileges—necessary to complete its task or function. For instance, if you're designing a smart thermostat, it should only have the access rights to control the temperature and not to your personal emails or bank details. By limiting the access rights of each IoT device, you can limit the potential damage if the device is compromised.

Principle 2: Defense in Depth

The Defense in Depth principle implies the use of multiple layers of security controls to protect your IoT systems. Imagine your IoT system as a castle. Now, you wouldn't rely solely on the outer wall to keep invaders out. You would have guards, a moat, inner walls, and so on. Similarly, for your IoT system, you could have firewalls, encryption, secure boot, intrusion detection systems, and regular software updates. This way, even if one layer is breached, others remain to safeguard your system.

Principle 3: Fail-Safe Defaults

The Fail-Safe Defaults principle means that, by default, access permissions should be set to deny, and the system should only allow access after the correct permissions have been verified. Let's take the example of a smart door lock. When you design this device, you should set its default state as 'locked'. Only when a valid key (physical or digital) is presented should it unlock. This way, even if the device fails or loses power, it defaults to its locked, secure state.

Principle 4: Separation of Duties

Separation of Duties involves splitting responsibility for critical actions among multiple entities to prevent fraud or error. For example, in an industrial IoT setting, one sensor might monitor temperature, another pressure, and a third one humidity. Each sensor only performs its duty and does not interfere with the others. This separation ensures that even if one sensor is compromised, the others remain functional and secure.

Principle 5: Security Updates and Patch Management

Regular updates and patches are crucial for maintaining security over the lifecycle of your IoT device. As new vulnerabilities are discovered, updates and patches can fix these issues before attackers can exploit them. For instance, if you're designing a smart camera, you should include a mechanism for securely updating the camera's software to address any new vulnerabilities.

Adopting 'Security by Design' greatly improves your IoT systems' security, recognizing it as a continuous process starting from the design phase.

Risk Assessment and Management in IoT

Understanding and managing risks in your IoT systems is as crucial as designing secure ones. This section guides you through risk assessment steps, introduces the concept of risk management, and demonstrates how to apply these principles to your IoT projects.

Risk Assessment in IoT

Risk assessment is the process of identifying potential threats, vulnerabilities, and risks associated with your IoT devices and systems. It's a crucial first step in managing the risks because you cannot effectively manage what you haven't identified or understood.

The risk assessment process generally involves the following steps:

- **Step 1: Asset Identification.** The first step is to identify what you need to protect. Your assets could be your IoT devices, the data they collect and process, the networks they connect to, and the services they provide. For example, if you're setting up a smart home, your assets might include smart thermostats, security cameras, the personal data they collect, and the home automation services they provide.

- **Step 2: Threat Identification.** Next, you need to identify potential threats to your assets. A threat could be anything that can exploit a vulnerability to cause harm to your assets. This could range from hackers trying to gain unauthorized access, to natural disasters like a fire or flood, to technical failures like power outages.

- **Step 3: Vulnerability Assessment.** This step involves identifying weaknesses in your IoT systems that could be exploited by threats. For example, you might find that your smart thermostat has a default password that's easy to guess, making it vulnerable to unauthorized access.

- **Step 4: Risk Evaluation.** Once you've identified your assets, threats, and vulnerabilities, the next step is to evaluate the risks. A risk is typically calculated as the potential impact of a threat exploiting a vulnerability, multiplied by the likelihood of that happening. For instance, if the impact of a hacker gaining access to your smart thermostat is high, and the likelihood is also high due to the default password vulnerability, then you have a high risk that needs to be addressed.

Risk Management in IoT

Once you've assessed your risks, the next step is risk management. This is the process of deciding how to handle each risk. You have several options:

- **Risk Mitigation.** Risk mitigation involves implementing controls to reduce the likelihood or impact of the risk. For instance, you could change the default password on your smart thermostat to a strong, unique password. This reduces the likelihood of a hacker being able to gain access.

- **Risk Transfer.** Risk transfer involves shifting the risk to another party. This is often done through insurance. For example, you might get an insurance policy that covers financial losses due to IoT security breaches.

- **Risk Acceptance.** Risk acceptance involves deciding to accept the risk as it is, usually because the cost of mitigating the risk would be higher than the potential loss. For instance, if the risk is that a hacker might change the temperature on your smart thermostat, you might decide to accept this risk if the potential harm is minor and the cost of mitigation is high.

- **Risk Avoidance.** Risk avoidance involves eliminating the risk entirely, often by not performing the activity that's associated with the risk. For example, if you determine that a particular brand of smart locks has significant security vulnerabilities and the manufacturer is not addressing them, you might choose not to use that brand of smart locks at all.

Effective risk management requires regular reviews and updates, as new threats emerge and old ones evolve. Understanding and applying these principles of risk assessment and management will allow you to significantly enhance the security and resilience of your IoT systems.

Incident Response and Disaster Recovery in IoT

Even with the best preventative measures in place, it's crucial to have a plan for when things go wrong. This section will guide you through the process of creating a comprehensive incident response plan and a robust disaster recovery strategy for your IoT systems.

Incident Response in IoT

Incident response is the organized approach to addressing and managing the aftermath of a security breach or cyberattack, also known as an incident. The goal is to handle the situation in a way that limits damage, reduces recovery time and costs, and improves the organization's resilience.

- **Step 1: Preparation.** Before an incident happens, you need to be prepared. This involves creating an incident response plan that outlines how to handle an incident, including roles and responsibilities, communication strategies, and procedures for analyzing and containing the incident.

  For example, if you detect unusual traffic from your IoT security camera, your response plan should guide you on the immediate steps to take, who to involve, and how to document the incident for future reference.

- **Step 2: Identification.** This step involves detecting and acknowledging that an incident has occurred. Effective detection requires continuous monitoring of your IoT systems. For instance, an anomaly in the power usage pattern of your smart fridge might indicate that it has been compromised.

- **Step 3: Containment.** Once you've identified an incident, the next step is to contain it to prevent further damage. This might involve disconnecting affected devices from the network or shutting down certain system functionalities.
- **Step 4: Eradication.** After containment, the next step is to find and eliminate the root cause of the incident. For your smart fridge example, this might involve removing the malicious software that's causing the power usage anomaly.
- **Step 5: Recovery.** Once the threat has been eradicated, you can begin to restore your systems to normal operation. For your smart fridge, this might involve reinstalling the device firmware and reconnecting it to the network.
- **Step 6: Lessons Learned.** After the incident, it's important to review what happened, what you did well, what you could have done better, and how you can prevent similar incidents in the future. This could lead to updates in your incident response plan and security controls.

Disaster Recovery in IoT

Disaster recovery, on the other hand, is a subset of business continuity planning. It focuses on restoring an organization's critical functions after a disaster, which could range from natural disasters like floods or fires, to human-induced disasters like cyber-attacks or equipment failure.

Your disaster recovery plan for IoT should include the following:

- **Data Backup.** Regular and secure backups of your IoT data ensure that you can restore your data if the original copies are lost or compromised. For example, you could regularly back up the configuration settings and historical data from your smart home devices to a secure cloud storage service.

- **Redundancy.** Having redundant IoT devices and systems can help you maintain functionality if a device or system fails. For example, if you have a network of smart sensors monitoring temperature in a warehouse, you might install extra sensors in case one or more of them fail.

- **Failover Systems.** A failover system automatically takes over the functions of a system that fails. In an IoT context, this could involve automatically switching to a backup server if the primary server fails.

- **Testing.** Regular testing of your disaster recovery plan is crucial to ensure that it works when you need it. This might involve simulating a disaster scenario and going through your recovery procedures to identify any issues or areas for improvement.

Understanding and implementing robust incident response and disaster recovery strategies are crucial steps in managing your IoT infrastructure's resilience. These strategies prepare you for incidents and disasters, minimize the potential damage they can cause, and ensure that your systems can recover and return to normal operation as quickly as possible.

A typical real-world application of these principles could be in an industrial IoT setting, such as a manufacturing plant. Let's say that the plant has a network of IoT sensors monitoring machinery health. If a sensor detects a potential equipment failure, the incident response plan would guide the staff on the immediate steps to take. These might include isolating the affected machine, investigating the issue, fixing the problem if possible, or replacing the machine if necessary. The disaster recovery plan would ensure that the plant can continue to operate during this process, for instance, by switching production to other machines.

Another example might be a smart city application, such as an IoT-based traffic control system. If a cyberattack were to compromise this system, the incident response plan would guide the city's IT staff on how to respond. This might include identifying and isolating the affected systems, removing the threat, and restoring the systems to normal operation. The disaster recovery plan would ensure that the city can manage its traffic effectively during this process, for instance, by switching to a backup traffic control system or implementing manual traffic control measures.

In conclusion, having a robust incident response and disaster recovery plan is not an option but a necessity in today's interconnected IoT world. These plans form your safety net when things go wrong, providing a structured approach to manage, recover from, and learn from incidents and disasters. Understanding and implementing these strategies will allow you to significantly enhance the resilience and reliability of your IoT systems.

Understanding the Regulatory Landscape of IoT

As the world becomes more connected, understanding the rules and regulations that govern the use of IoT technology is becoming increasingly important. It will not only ensure your IoT applications and systems are legally compliant, but it can also guide your decisions and strategies as you navigate the IoT space.

IoT Regulations: An Overview

IoT regulations are laws and guidelines designed to address various issues related to IoT, such as privacy, data protection, security, and interoperability. These regulations can be specific to certain industries, types of data, or even specific types of IoT devices. They can be set by governmental bodies, international organizations, or industry groups.

Privacy and Data Protection Regulations

One of the most important areas of regulation in IoT involves privacy and data protection. With IoT devices often collecting vast amounts of personal data, it's essential to understand the regulations that govern how this data can be collected, stored, processed, and shared.

For example, if you're operating in the European Union, you'll need to comply with the General Data Protection Regulation (GDPR). This law has strict rules about consent for data collection, individuals' rights to access and control their data, and data breach notification requirements.

In the U.S., there's no single, comprehensive federal law regulating the collection and use of personal data. Instead, various federal and state laws apply, such as the California Consumer Privacy Act (CCPA).

Security Regulations

There are also regulations specifically focused on IoT security. These typically aim to ensure that IoT devices and systems have adequate security measures in place to protect against threats.

For instance, California's SB-327 law, often referred to as the first IoT-specific security law in the U.S., requires manufacturers to equip IoT devices with "reasonable" security features. Depending on the type of IoT device you're developing or using, you may also need to comply with industry-specific security standards, such as the Health Insurance Portability and Accountability Act (HIPAA) for healthcare IoT devices.

Interoperability Regulations

Interoperability - the ability of different IoT systems and devices to work together - is another important area of regulation. Standards and regulations around interoperability are crucial for enabling the broad range of IoT applications we see today.

For example, if you're developing an IoT solution for smart homes, your devices might need to be compatible with certain communication protocols or platforms, such as Zigbee or Apple's HomeKit, which have their own requirements and standards.

Navigating the Regulatory Landscape

Given the complexity and diversity of the regulatory landscape, it's crucial to understand the regulations that apply to your specific IoT applications:

- **Stay Informed:** Keep up-to-date with the latest regulatory developments in your industry and the regions where you operate. This might involve subscribing to regulatory news updates, attending industry conferences, or joining industry groups.

- **Seek Legal Advice:** Given the complexities of IoT regulations, it's often a good idea to seek legal advice.

Lawyers who specialize in IoT, data protection, or your specific industry can provide valuable guidance.

- **Build Compliance Into Your Processes:** Rather than treating compliance as an afterthought, build it into your IoT development and deployment processes. This "compliance by design" approach can save you time and effort in the long run.
- **Document Your Compliance Efforts:** Keep detailed records of how you're complying with regulations. This documentation can be invaluable if you ever need to demonstrate your compliance to regulators, customers, or other stakeholders.

Understanding the regulatory landscape of IoT will allow you to ensure that your IoT applications are compliant, enhance the trust of your users, and ultimately, drive the success of your IoT projects.

Module 5 Exam Questions

1. What kind of approach is 'Security by Design' in IoT?

a. An approach focusing on designing attractive IoT devices

b) An approach that incorporates security principles from the initial design phase

c) An approach that considers security as a post-deployment activity

d) An approach that ignores security considerations

2. In the context of IoT, what does "Risk Assessment" involve?

a) Ignoring potential risks

b) Designing risk-free devices

c) Focusing only on device performance

d) Identifying and evaluating potential security threats

3. What is a crucial factor to consider when designing an IoT system?

a) Ensuring built-in security from the initial stages

b) Focusing solely on device functionality

c) Prioritizing aesthetics over security

d) Ignoring potential security risks

4. What is an essential part of an effective IoT security strategy?

a) Ignoring regulatory compliance

b) Assuming security risks don't exist

c) Implementing an incident response plan

d) Prioritizing system design over security

5. What is the biggest security challenge in IoT?

a) Limited computational power

b) Data privacy issues

c) Both a and b

d) None of the above

6. What is the primary focus of IoT security best practices?

a) To ensure data privacy and integrity

b) To enhance device performance

c) To improve user interfaces

d) None of the above

7. In the context of IoT, what does encryption do?

a) Increases the speed of data transmission

b) Changes data into a code to prevent unauthorized access

c) Helps to connect to the internet

d) None of the above

8. What is the primary concern of data privacy and compliance in IoT?

a) Ensuring data is transmitted fast

b) Ensuring data is stored securely and used responsibly

c) Ensuring data is colorful

d) Ensuring data is loud

9. Which one of the following is not an area of IoT application?

a) Smart Homes and Buildings

b) Industrial IoT (IIoT) and Manufacturing

c) Fashion Designing

d) Healthcare and IoT

10. What is the emerging trend in IoT technology?

a) Less data generation

b) Decreased connectivity

c) Increased security concerns

d) All of the above

11. What does Security by Design imply in the context of IoT?

a) Devices designed without any security features

b) Incorporating security measures in the initial design phase of IoT devices

c) Security measures are optional for IoT devices

d) Security measures can be added later without any planning

12. What does Risk Assessment in IoT involve?

a) Ignoring potential threats

b) Identifying and evaluating potential risks and vulnerabilities

c) Assessing only the financial risks

d) Assessing only the operational risks

13. What is the primary focus of Incident Response in IoT?

a) Enhancing device performance

b) Managing and recovering from security incidents

c) Increasing data transmission speed

d) None of the above

14. What does the General Data Protection Regulation (GDPR) regulate?

a) Data privacy and protection in the European Union

b) Data transmission speed

c) IoT device manufacturing standards

d) None of the above

15. What is the primary concern of the Health Insurance Portability and Accountability Act (HIPAA) in terms of IoT?

a) Ensuring fast data transmission

b) Ensuring colorful data

c) Ensuring secure and private health data

d) None of the above

16. What does Interoperability in IoT imply?

a) Devices working independently

b) Devices working together

c) Devices working rarely

d) Devices not working at all

17. What does the term 'compliance by design' mean in the context of IoT?

a) Compliance is an optional feature

b) Compliance is built into the development and deployment processes

c) Compliance is a post-deployment activity

d) Compliance is not necessary for IoT devices

18. What is the primary focus of California's SB-327 law?

a) IoT device manufacturing

b) IoT device security

c) IoT device sales

d) IoT device design

19. What kind of IoT regulations are designed to ensure that IoT devices and systems have adequate security measures in place?

a) Interoperability regulations

b) Security regulations

c) Privacy and data protection regulations

d) None of the above

20. Which IoT application area involves the use of IoT technology in the production and manufacturing processes?

a) Smart Homes and Buildings

b) Healthcare and IoT

c) Industrial IoT (IIoT) and Manufacturing

d) Smart Cities and Infrastructure

Answers:

1. b
2. d
3. a
4. c
5. c
6. a
7. b
8. b
9. c
10. c
11. b
12. b
13. b
14. a
15. c
16. b
17. b
18. b
19. b
20. c

Module 6. IoT Connectivity and Networking

Introduction to IoT Connectivity and Networking

Welcome to Module 6 on IoT Connectivity and Networking. In this module, we will embark on an exciting exploration of the complex, yet fascinating world of connectivity and networking within the Internet of Things (IoT) ecosystem.

The Internet of Things, as you may already know, refers to the vast network of interconnected devices and sensors, all of which communicate and exchange data to fulfill a myriad of functions. From everyday devices such as smartphones and smart appliances to industrial equipment like sensors on an assembly line, IoT is transforming how we live, work, and interact with the world around us. And at the heart of this transformation lies the critical aspect of connectivity and networking.

Connectivity and networking are the lifelines of IoT systems, creating the necessary bridges for data to travel and enabling devices to interact in meaningful ways. Without robust connectivity and efficient networking, the 'Internet' part of the 'Internet of Things' would simply not exist.

In this module, you will learn about the crucial role that connectivity and networking play in IoT. We will delve into the various technologies and protocols that make IoT connectivity possible, exploring everything from wireless technologies to low-power wide-area networks (LPWAN). We will examine the principles and strategies behind network topologies and design, and the importance of effective network management and optimization for IoT systems.

To bring these concepts to life, we will be using practical examples from real-world applications of IoT. Imagine a smart city where thousands of sensors collect data to optimize traffic flow, energy usage, and public safety. Or consider a manufacturing plant where machinery communicates in real time to maximize efficiency and minimize downtime. These are just some of the scenarios we will delve into as we explore the world of IoT connectivity and networking.

This module is not just about understanding the theoretical aspects of IoT networking; it is also about appreciating its practical implications and future potential. As we navigate through this journey, we hope to equip you with the knowledge and insights you need to understand, and perhaps even contribute to, the ever-evolving landscape of IoT connectivity.

So, let's embark on this exciting journey together.

Wireless Technologies for IoT

Wireless technologies are the lifeblood of IoT, forming the backbone that enables different devices to connect and communicate without the need for physical wires. These technologies are diverse, each with its unique attributes, strengths, and trade-offs. It's essential to select the appropriate wireless technology for your IoT project, taking into account various factors like range, data requirements, power consumption, and cost.

Wi-Fi

Let's start with the most common and widely known wireless technology: Wi-Fi. Wi-Fi is a crucial enabler for IoT applications, especially those in a home or office environment. Your smart thermostat, for instance, uses Wi-Fi to send temperature data to your smartphone, allowing you to control your home's heating remotely.

Wi-Fi's key advantages include its high data rates and large coverage area, typically up to 100 meters. However, Wi-Fi devices can be power-hungry, which can be a limitation for battery-powered IoT devices. Furthermore, Wi-Fi operates in the crowded 2.4 GHz and 5 GHz bands, which can lead to interference in densely populated environments.

Bluetooth and Bluetooth Low Energy (BLE)

Next on our list are Bluetooth and its energy-efficient variant, Bluetooth Low Energy (BLE). Bluetooth is ideal for short-range, point-to-point, and small network communications. Think of your fitness tracker syncing data with your smartphone – that's Bluetooth in action. BLE, on the other hand, is optimized for low power consumption, making it an excellent choice for battery-powered devices that send small amounts of data.

Zigbee and Z-Wave

Zigbee and Z-Wave are wireless technologies specifically designed for low-cost, low-power IoT applications. They operate on the mesh network principle, where each device (or 'node') in the network can relay data to other nodes. This functionality extends the range of the network and provides redundancy – if one device fails, others can still communicate. Zigbee and Z-Wave are commonly used in smart home applications – for example, controlling your lights, locks, and security systems.

Cellular IoT

Cellular IoT technologies leverage the existing cellular network infrastructure to provide wide-area coverage and high reliability. They are ideal for IoT applications that require mobility or are deployed in remote locations. Examples of Cellular IoT technologies include NB-IoT (Narrowband IoT) and LTE-M (Long Term Evolution for Machines). Imagine a fleet of delivery trucks fitted with IoT devices for real-time tracking – this is a typical application where Cellular IoT shines.

Low-Power Wide-Area Networks (LPWAN)

Finally, we have Low-Power Wide-Area Networks (LPWAN) – a group of technologies designed for long-range communication and minimal power consumption. LPWANs are ideal for applications where devices need to send small amounts of data over long distances while conserving battery life. LPWAN technologies include LoRaWAN (Long Range WAN), Sigfox, and NB-IoT. An example application could be a city-wide network of environmental sensors measuring air quality.

Each of these wireless technologies has its strengths and limitations. Therefore, the choice of technology depends on your specific IoT application. You need to consider several factors, including the amount of data to be transmitted, the required range, power constraints, cost, and the need for mobility or fixed installation.

In the next sections, we'll delve deeper into each of these technologies, looking at how they work underneath the hood, their specific use-cases, and how they fit into the larger IoT ecosystem.

Wi-Fi Deep Dive

Your experience with Wi-Fi probably starts and ends with connecting your devices to the internet, but there's much more to Wi-Fi than meets the eye, especially in an IoT context. Wi-Fi is a versatile and powerful communication technology that operates in the 2.4 GHz and 5 GHz bands, allowing for fast data transmission rates, often up to several hundred Mbps.

For IoT applications that need to transfer large amounts of data, such as video streaming for a security camera or data-intensive industrial monitoring systems, Wi-Fi is often the go-to technology. However, one aspect to keep in mind is that Wi-Fi can consume a significant amount of power, which might not make it the ideal choice for devices running on batteries that need to last for extended periods without replacement.

Bluetooth and BLE - Beyond Pairing Devices

The Bluetooth technology, especially BLE, has found widespread use in IoT applications. BLE is designed for short bursts of long-range radio connections, making it perfect for applications that do not require continuous connectivity but depend on long battery life.

Think of a heart rate monitor or a smartwatch; they use BLE to transmit data to your smartphone at regular intervals, consuming far less power than traditional Bluetooth or Wi-Fi connections. BLE operates in the 2.4 GHz ISM band and can achieve a range of up to 100 meters in ideal conditions.

Zigbee and Z-Wave - Creating IoT Mesh Networks

Zigbee and Z-Wave are particularly interesting because they can create a mesh network. In a mesh network, each device is interconnected with others, allowing data packets to hop from one device to another until they reach their destination. This not only increases the effective range of the network but also provides redundancy – if one device fails, the network can self-heal by routing data through other devices.

Imagine having a network of Zigbee-enabled smart light bulbs in your home. If the nearest light bulb to the hub fails, the hub can still control the other bulbs by routing commands through the functioning bulbs. That's the power of a mesh network!

Cellular IoT - Utilizing Existing Infrastructure

Cellular IoT brings the advantages of existing cellular networks to the IoT – broad coverage, high reliability, and mobility. Cellular IoT technologies, like NB-IoT and LTE-M, are designed for devices that need to transmit small amounts of data over long distances. These technologies are power-efficient and can penetrate hard-to-reach indoor locations.

For example, a logistics company could equip its containers with cellular IoT devices to monitor the containers' location and condition, no matter where they are in the world. Cellular IoT can also be used to connect rural IoT devices, like agricultural sensors, where other types of connectivity might not be feasible.

Low-Power Wide-Area Networks (LPWAN) - For City-Wide or Rural Applications

LPWAN technologies are specifically designed for long-range, low-power IoT applications. If your IoT application involves devices spread out over a city or in rural areas, and these devices only need to send small amounts of data infrequently, then LPWAN technologies like LoRaWAN or Sigfox could be the ideal choice.

For instance, a city could use LoRaWAN to connect different environmental sensors spread across the city, measuring parameters like air quality, temperature, or noise levels. Since these sensors only need to send small amounts of data a few times per hour, and they need to last for years on a single battery, LPWAN is the perfect fit.

As you can see, the world of wireless technologies for IoT is vast and varied. The key to successful IoT implementation is understanding these technologies and choosing the right one based on the specific requirements of your IoT application.

Factors to Consider When Choosing a Wireless Technology

When you're selecting a wireless technology for your IoT project, it's important to understand the trade-offs involved. Here are a few factors to consider:

- **Range:** This is the maximum distance between your IoT devices and the gateway or hub that can still allow for reliable communication. Wi-Fi and cellular technologies offer a large range, while Bluetooth and BLE are more suitable for short-range applications. LPWAN technologies like LoRaWAN offer a good middle ground with several kilometers of range.

- **Data Rate:** This is the volume of data your devices need to transmit within a given time. If you're working on an application like a smart security camera that needs to send high-definition video, Wi-Fi might be the best choice due to its high data rate. But if your devices only need to send small amounts of data, like temperature or humidity readings, a lower data rate technology like Zigbee or LoRaWAN would be more appropriate.

- **Power Consumption:** If your IoT devices are battery-powered and need to operate for a long time without a battery replacement, power consumption becomes a critical factor. Technologies like BLE, Zigbee, and LPWAN are designed to be power-efficient, making them suitable for such applications.

- **Cost:** The cost of the IoT devices themselves, as well as the cost of the infrastructure, can also influence your choice of wireless technology. For example, cellular IoT might involve ongoing subscription costs, while a Zigbee network might require a more expensive gateway device.

- **Environment:** The environment where your IoT devices will be deployed can also affect your choice. For example, Wi-Fi and Bluetooth might not perform well in environments with lots of interference, while cellular IoT can work virtually anywhere there's a cellular signal.

As we delve deeper into the world of IoT connectivity and networking, you'll gain a better understanding of these factors and how to balance them according to your IoT project's specific needs.

In the next sections, we'll discuss more advanced topics, like network protocols, topologies, and optimization techniques. We'll also cover the important aspect of network security and how to protect your IoT devices from potential threats. But for now, you should have a solid foundation in the various wireless technologies for IoT and how to select the right one for your project. It's an exciting journey, and we're just getting started!

Low-Power Wide-Area Networks (LPWAN)

In this section, we'll delve deeper into LPWAN, a critical wireless technology category for IoT. By the end of this section, you'll have a comprehensive understanding of what LPWAN is, its key characteristics, the major types of LPWAN technologies, and how they can be used in various IoT applications.

Understanding LPWAN

LPWAN stands for Low-Power Wide-Area Network. As the name suggests, LPWAN technologies are designed to provide long-range communication between devices while consuming very little power. This makes them perfect for IoT applications where devices are spread out over a large area and need to run for long periods (often several years) on a single battery.

LPWAN technologies are optimized for transmitting small amounts of data over long distances. They operate at low data rates, usually under a few kbps, which is sufficient for many IoT applications that only need to send small packets of data infrequently. This could include applications like environmental monitoring, asset tracking, or smart metering, where devices might only need to send a few bytes of data a few times per day.

Major Types of LPWAN Technologies

There are several types of LPWAN technologies, each with its own strengths and trade-offs. Here, we'll focus on the three most common ones: LoRaWAN, Sigfox, and NB-IoT.

- **LoRaWAN:** LoRaWAN stands for Long Range Wide Area Network. It's a protocol designed for wireless battery-operated devices in a regional, national, or global network. LoRaWAN offers features like low power consumption, long-range capability (up to 15 km in rural areas), and secure data transmission. It's ideal for applications where devices need to send small amounts of data over long distances, like smart agriculture or environmental monitoring.

- **Sigfox:** Sigfox is another LPWAN technology that provides a simple, efficient, and scalable solution for IoT devices to communicate. Sigfox operates in the sub-GHz frequency bands and is designed for small, infrequent data transmissions, up to a maximum of 140 messages per day. It's used in applications like asset tracking, smart metering, and security systems.
- **NB-IoT:** NB-IoT, or Narrowband IoT, is a cellular-based LPWAN technology. It leverages existing cellular infrastructure to provide excellent coverage, including hard-to-reach indoor and underground locations. NB-IoT is well-suited for applications requiring reliable, long-range communication, and low power consumption, like smart city projects or industrial IoT.

Practical Applications of LPWAN

To better understand LPWAN technologies and their uses, let's look at some practical examples.

Suppose you're working on a smart farming project where you need to monitor soil moisture levels across a large farm. Your sensors need to send small amounts of data (the moisture readings) over several kilometers (the size of the farm) and need to run for years on a single battery (to reduce maintenance costs). In this case, an LPWAN technology like LoRaWAN would be an excellent choice. LoRaWAN's long range and low power consumption make it perfect for this application.

Now, imagine you're developing a smart city application where you need to monitor air quality at various locations throughout the city. Your devices need to send data over a city-wide area and penetrate buildings to reach some indoor locations. Here, NB-IoT would be a great choice because it can leverage existing cellular infrastructure to provide city-wide coverage and excellent building penetration.

In another scenario, you might be running a logistics company and want to keep track of your assets as they travel across the country or even around the world. Sigfox, with its global coverage and focus on small, infrequent data transmissions, would be a perfect fit for this application. With Sigfox, you could install small, battery-powered trackers on your assets that send their location and status a few times a day, providing you with real-time tracking and monitoring capabilities.

Choosing the Right LPWAN Technology

Just as with any technology decision, the choice of which LPWAN technology to use depends on the specific requirements of your IoT project. Here are a few factors you should consider when making your decision:

- **Range:** All LPWAN technologies offer long-range communication, but the exact range can vary based on the technology and the environment. For example, LoRaWAN can provide up to 15 km of range in rural areas, but in urban areas with more interference, the range might be reduced.

- **Data Rate:** LPWAN technologies are designed for low data rate applications, but there can still be differences between the technologies. For example, NB-IoT can support higher data rates than LoRaWAN or Sigfox, making it a better choice for applications that need to send slightly larger amounts of data.

- **Power Consumption:** All LPWAN technologies are designed to be power-efficient, but there can be differences based on the technology and how often your devices need to send data. For example, Sigfox is designed for devices that only need to send data a few times per day and can be extremely power-efficient in these scenarios.

- **Coverage:** If your IoT application needs to operate over a city-wide, country-wide, or even global area, you'll need to consider the coverage provided by the LPWAN technology. NB-IoT leverages existing cellular infrastructure, so it can provide excellent coverage in areas with good cellular reception. On the other hand, Sigfox operates its own global network, allowing for seamless coverage in over 70 countries.
- **Cost:** The cost of the LPWAN technology can include the cost of the devices themselves, as well as any network usage fees. Some LPWAN technologies, like LoRaWAN, can be operated on private networks, allowing you to avoid ongoing network usage fees. Others, like NB-IoT and Sigfox, might involve ongoing subscription costs.

In the end, the choice of LPWAN technology comes down to understanding your IoT project's specific requirements and how each technology can meet those requirements. The power of LPWAN lies in its ability to provide long-range, power-efficient connectivity for a wide range of IoT applications, and understanding this technology is a crucial step on your journey into the world of IoT.

Next, we'll dive into another vital aspect of IoT connectivity: network topologies and design. Stay tuned!

Network Topologies and Design

In this section we'll explore the various ways in which devices in an IoT system can be organized or interconnected. By the end of this, you will have a comprehensive understanding of different network topologies, their strengths, and weaknesses, and how to design and select the appropriate topology for your IoT project.

Understanding Network Topologies

A network topology refers to the arrangement of various elements (links, nodes, etc.) of a computer network. Essentially, it is the topological structure of a network and may be depicted physically or logically. In the context of IoT, the network topology represents how your devices (sensors, actuators, gateways, etc.) are connected to each other and to the internet. The choice of network topology can have a significant impact on the performance, cost, and reliability of your IoT system.

There are several types of network topologies, each with its own advantages and disadvantages. Here are some of the most common ones:

- **Star Topology:** In a star topology, all devices are connected to a central node or hub, like the spokes of a wheel. The hub could be a gateway device that collects data from the other devices and connects to the internet. A star topology is simple to set up and manage, but it relies heavily on the hub. If the hub fails, all communication is disrupted.

- **Mesh Topology:** In a mesh topology, devices are interconnected, with each device connected to multiple others. This creates multiple paths for data to travel, improving reliability. If one device fails, data can still reach its destination via a different path. Mesh networks are robust and offer excellent coverage but can be more complex to set up and manage.

- **Tree Topology:** A tree topology is a variation of the star topology, where multiple star networks are connected to a central 'root' node. This allows for a large number of devices to be connected and is commonly used in Zigbee networks.

- **Bus Topology:** In a bus topology, all devices are connected to a common backbone or bus. Data from any device is available to all others on the bus. Bus topologies are simple and cheap to implement but can suffer from data collisions if many devices try to send data at the same time.

Practical Applications of Network Topologies

To illustrate the use of network topologies in real-world IoT applications, let's consider a few examples.

Imagine you're setting up a smart home system where you have several smart devices (lights, thermostats, security cameras, etc.) connected to a central hub, which is connected to the internet. This is an example of a star topology. The hub acts as the central node, collecting data from the devices and controlling them based on your instructions. If you were to add a smart light bulb to your system, you would connect it to the hub, and it would immediately become part of your smart home network.

Now consider an industrial IoT application where you have hundreds of sensors spread out across a large factory floor. Each sensor monitors a machine's performance and needs to send its data back to a central server. Instead of connecting each sensor to the server (which would require a lot of wiring), you could set up a mesh network. Each sensor could connect to a few nearby sensors, creating multiple paths for data to travel back to the server. If a sensor fails or a path is blocked, the data can simply take a different path.

Designing Your IoT Network

When designing your IoT network, there are several factors to consider when choosing a topology:

- **Size and Scalability:** How many devices do you need to connect, and how might this number change in the future?
- **Reliability:** How crucial is it that your network stays up and running? If a single device fails, can the network adjust and keep the data flowing?
- **Coverage:** How far apart are your devices? Do you need to cover a large area, or are all your devices close together?
- **Power Consumption:** Some topologies require devices to transmit data over long distances or to constantly maintain multiple connections, which can consume more power.
- **Cost:** Different topologies require different types of devices and infrastructure, which can affect the cost. For example, a mesh network might require more sophisticated devices that can maintain multiple connections, which could be more expensive than the simple devices used in a star network.

In a star topology, if you're planning to deploy devices in a large area or need to ensure high reliability, a mesh network might be a better choice. A tree topology could be suitable for hierarchical structures, like a multi-floor building where each floor could form a star network connected to the central 'root' node (perhaps placed in a central control room). A bus topology might be suitable for a line of sensors along a pipeline or a conveyor belt.

Remember, the choice of network topology will significantly influence the network's performance, the cost of deployment, and how you manage and maintain the network. Therefore, taking the time to understand different network topologies and thoughtfully designing your IoT network is a crucial step in building a successful IoT system.

In the next section, we will explore another important aspect of IoT connectivity: Network management and optimization. As your IoT system grows and evolves, effectively managing your network and optimizing its performance will become increasingly important. So, make sure to join us in the next section as we continue our journey into the world of IoT connectivity and networking.

IoT Network Management and Optimization

As the complexity and scale of your IoT network increases, managing the network and ensuring it operates optimally becomes increasingly essential. By the end of this section, you will understand the various challenges and strategies involved in managing and optimizing an IoT network.

Understanding IoT Network Management

IoT network management involves a range of activities aimed at ensuring the network's smooth operation. These can include:

- **Device Management:** This involves monitoring and maintaining the devices on your network, such as sensors, actuators, and gateway devices. Tasks can include configuring devices, updating firmware, diagnosing and fixing faults, and managing the device lifecycle.

- **Performance Management:** This involves monitoring and optimizing the performance of your network. Key performance indicators (KPIs) might include data transmission rates, latency, packet loss rate, and uptime.

- **Security Management:** Security is a critical concern in IoT networks. Security management includes activities like setting and enforcing security policies, monitoring for suspicious activity, managing encryption keys, and responding to security incidents.

- **Data Management:** Your IoT devices generate potentially vast amounts of data, which need to be reliably and efficiently collected, stored, processed, and transmitted.

Challenges in IoT Network Management

IoT network management can pose a number of challenges:

- **Scale:** IoT networks can include thousands or even millions of devices, making management tasks complex and time-consuming.
- **Diversity:** IoT networks can include a wide range of devices from different manufacturers, each with its own capabilities, requirements, and management interfaces.
- **Connectivity:** IoT devices can be spread over a large area and might use various connectivity technologies, making it challenging to maintain reliable connections and manage the network.
- **Security:** IoT devices can be vulnerable to various security threats, from device tampering to data interception, making security management a critical task.

Practical Strategies for IoT Network Management and Optimization

Given these challenges, how can you effectively manage and optimize your IoT network? Let's discuss some practical strategies.

- **Automated Device Management:** To handle the scale and diversity of devices on your network, consider using an IoT platform that provides automated device management features. For example, you might use a platform that allows you to define device profiles, which specify the configuration settings for each type of device on your network. When a new device is added to the network, the platform can automatically apply the correct profile, saving you the effort of manually configuring the device.
- **Remote Monitoring and Diagnostics:** To manage devices spread over a large area, use an IoT platform that supports remote monitoring and diagnostics. For example, if a device on your network is not performing correctly, you can use the platform to remotely access the device, check

its status, reset it, or even update its firmware, all without needing to physically visit the device's location.

- **Network Performance Monitoring:** To optimize your network's performance, use tools that provide real-time monitoring of key performance indicators. For example, you might use a network monitoring tool that shows you the current data transmission rate, latency, and packet loss rate for each device on your network. If performance starts to degrade, you can quickly identify the problem and take corrective action.

- **Security Management Tools:** To protect your network against security threats, use an IoT platform that provides robust security management tools. This might include features for managing encryption keys, setting and enforcing security policies, and detecting and responding to security incidents.

- **Data Optimization:** Given the potentially large amounts of data generated by your IoT devices, consider strategies for optimizing data transmission and storage. For example, you might employ edge computing strategies, where data processing is done near the source of data generation, reducing the amount of data that needs to be transmitted. Or you might use data compression techniques to reduce the amount of storage space required.

Practical Examples of IoT Network Management and Optimization

To illustrate these concepts, let's consider a practical example. Suppose you're managing an IoT network for a smart city project, with thousands of sensors spread across the city monitoring various aspects like traffic, air quality, and energy usage.

In such a scenario, automated device management would be crucial. With a diverse array of sensors from various manufacturers, manually configuring each one would be a herculean task. Using an IoT platform that enables you to define device profiles and automatically apply them to new devices as they are added to the network would save you considerable time and effort.

Furthermore, given the wide distribution of devices, remote monitoring and diagnostics would be invaluable. If a traffic sensor on a busy intersection stops transmitting data, you could use your IoT platform to remotely check the device's status, reset it, or update its firmware, all without disrupting traffic to physically access the device.

In this scenario, network performance monitoring is also critical. You might set up a dashboard that displays real-time data on network performance, showing the data transmission rate, latency, and packet loss rate for each sensor. If a sensor starts experiencing high latency or packet loss, you could quickly identify the issue and take corrective action - perhaps adjusting the network topology or adding a new gateway device to improve connectivity.

And of course, given the sensitive nature of some data (like air quality data that could influence public health policies), robust security management would be a must. You would need an IoT platform with strong security features, allowing you to manage encryption keys, enforce security policies, and swiftly respond to any security incidents.

Finally, with thousands of sensors all transmitting data, data optimization strategies would be crucial. You might employ edge computing techniques, with gateway devices across the city processing data locally and only transmitting summarized data to the cloud. This would reduce the load on your network and minimize data transmission costs.

In conclusion, while managing and optimizing an IoT network involves complexity, appropriate strategies can ensure reliability and efficiency. Up next, we'll explore IoT platforms and ecosystems that offer useful tools for these tasks.

Interoperability in IoT Networking

Interoperability is an essential consideration in your IoT networking journey. This section will unpack the concept of interoperability, discuss why it matters in IoT, explore the challenges it presents, and provide practical strategies for achieving it.

Understanding Interoperability

In the simplest terms, interoperability is the ability of different systems, devices, applications, or products to work together, or 'inter-operate'. In the context of IoT, interoperability is the ability of your devices, from sensors to gateways, and systems, from data storage to user interfaces, to work together seamlessly.

Interoperability is crucial for several reasons:

- **Device Diversity:** Your IoT network might include a wide range of devices from different manufacturers, each with its own specifications, protocols, and interfaces. Interoperability ensures that these diverse devices can communicate and work together.

- **System Integration:** Your IoT network doesn't exist in isolation. It needs to integrate with other systems, such as data storage systems, data processing systems, and user interface systems. Interoperability is vital to achieve this integration.

- **Scalability:** As your IoT project grows, you'll likely add new devices and systems to your network. Interoperability ensures that these new components can work with the existing ones.

Challenges in Achieving Interoperability

Achieving interoperability in IoT can be challenging due to:

- **Standardization:** There are many different standards in IoT, from communication protocols to data formats. Ensuring that all your devices and systems adhere to the same standards can be difficult.

- **Legacy Systems:** You might need to integrate your IoT network with existing legacy systems, which may not be designed with IoT or interoperability in mind.

- **Security:** Interoperability can introduce security risks. For example, if you're integrating with a system that has weaker security measures, it could potentially expose your IoT network to threats.

Practical Strategies for Achieving Interoperability

Despite these challenges, there are several practical strategies you can use to achieve interoperability in your IoT network:

- **Adopt Open Standards:** Wherever possible, use devices and systems that adhere to open standards. Open standards are publicly available and have been agreed upon by the industry, which makes them widely supported and helps ensure interoperability.

- **Use an IoT Platform:** Many IoT platforms are designed with interoperability in mind. They can support a wide range of devices, protocols, and interfaces, and offer features to help integrate with other systems.

- **Implement Middleware:** In situations where you're dealing with devices or systems that don't naturally interoperate, consider implementing middleware. Middleware is software that sits between your devices and systems and helps them communicate by translating between different protocols and formats.

Practical Examples of Interoperability in IoT Networking

Now, let's consider a practical example to illustrate these concepts. Suppose you're developing a smart building system, with IoT devices for controlling lighting, heating, cooling, and security.

In this scenario, you might have lighting controllers from one manufacturer, temperature sensors from another, and security cameras from yet another. Each of these devices might use different communication protocols and data formats. Here, adopting open standards could significantly help. By choosing devices that all support the same standard, like MQTT for communication and JSON for data formatting, you ensure that these diverse devices can all communicate seamlessly.

Next, integrating these IoT devices with the building's existing systems, like the HVAC system and the security monitoring system, could be a challenge. This is where an IoT platform could be beneficial. You might choose a platform that supports a wide range of protocols and interfaces, and offers integration features, like APIs and SDKs, which make it easier to connect your IoT devices with these systems.

However, suppose the building's HVAC system is an older system that doesn't support the MQTT protocol your devices use. This is where middleware could come in handy. You could implement a middleware solution that takes the MQTT messages from your devices and translates them into a protocol the HVAC system can understand.

Now, let's imagine you want to add a new type of device to your network, like smart blinds that adjust based on the sunlight level. Interoperability is crucial here to ensure that these new devices can work with the existing ones. By sticking to open standards and using an IoT platform that supports a wide range of devices, you can add these new devices to your network with minimal fuss.

Finally, throughout this process, you need to be mindful of security. Each new device and system you add to your network could potentially introduce new vulnerabilities. It's important to choose devices with robust security features, use an IoT platform that offers strong security controls, and regularly review and update your security measures to protect your network.

In conclusion, interoperability is a complex but crucial aspect of IoT networking. Understanding its importance, being aware of the challenges, and employing strategies like adopting open standards, using an IoT platform, and implementing middleware, will allow you to build an IoT network where all components work together seamlessly, making your smart building project a success. As we move to the next module, we'll continue to explore how you can effectively use and manage the various tools and platforms available in the world of IoT.

Quality of Service (QoS) in IoT Networking

Quality of Service, often abbreviated as QoS, is a pivotal concept in the realm of networking, including the Internet of Things. In this section, you will learn about QoS, its importance in IoT networking, the challenges associated with it, and how to effectively implement it in your network.

Understanding Quality of Service

Quality of Service refers to the performance level of a service or a network. In IoT networking, QoS is about managing your network's resources to provide a satisfactory level of service, ensuring efficient communication between your IoT devices.

QoS encompasses several parameters, including:

- **Latency:** The time it takes for a packet of data to travel from source to destination.
- **Jitter:** The variability in latency.
- **Bandwidth:** The maximum data transfer rate of a network.
- **Packet Loss:** The percentage of packets that are sent but never reach their destination.

These parameters can significantly impact the performance of your IoT applications. For example, a high latency or packet loss rate could make a real-time monitoring application unusable, while insufficient bandwidth could prevent your devices from transmitting data efficiently.

Importance of QoS in IoT Networking

In an IoT network, you could have numerous devices, each with different communication needs. For example, a real-time health monitoring device might require low latency and minimal packet loss, while a weather sensor that sends data only once an hour might be more flexible on these parameters but need minimal bandwidth.

QoS is crucial here to prioritize network resources based on the needs of your devices and applications. Without effective QoS management, your network could become congested, leading to poor performance and unsatisfactory service.

Challenges in Implementing QoS

Implementing QoS in your IoT network can be challenging due to:

- **Device Diversity:** Your IoT network might include a wide range of devices with diverse communication needs. Catering to all these needs can be complex.
- **Network Dynamics:** The state of your network can change rapidly, with devices joining or leaving, data traffic fluctuating, and network conditions varying.
- **Scalability:** As your IoT project grows, managing QoS for an increasing number of devices can become increasingly complex.

Practical Strategies for Implementing QoS

Despite these challenges, there are several strategies you can use to manage QoS in your IoT network:

- **Prioritize Traffic:** Not all data traffic in your IoT network is equally important. By identifying and prioritizing critical traffic, you can ensure that your most important applications always receive the necessary network resources.
- **Monitor Network Performance:** Regularly monitoring your network's performance can help you identify issues like congestion or packet loss and take corrective action.

- **Use QoS Tools:** Many networking devices and IoT platforms offer QoS features that can help you manage network resources. These tools can automatically prioritize traffic, limit bandwidth usage, and adjust parameters to maintain a satisfactory level of service.

Practical Examples of QoS in IoT Networking

To illustrate these concepts, let's consider a practical example. Suppose you're managing an IoT network for a smart factory, with devices monitoring machinery in real-time, tracking inventory, and measuring environmental conditions.

In this scenario, the real-time machinery monitors might be your most critical devices. These devices need to transmit data with low latency and minimal packet loss to enable instant response if a problem is detected. Here, implementing traffic prioritization could be crucial. You could configure your network to always prioritize traffic from these devices, ensuring they always have sufficient network resources.

To handle network dynamics, you might implement a network performance monitoring system. This system could provide real-time data on network parameters like latency, jitter, and packet loss. With this information, you can proactively detect and address network performance issues, ensuring a seamless operation of your IoT devices. There are various network monitoring tools available today, both open-source and commercial, that can provide these insights.

Additionally, you might consider the integration of machine learning algorithms to predict network behavior and to automate the process of network optimization. Machine learning can process huge amounts of network data to identify patterns and predict potential network issues before they occur. For example, if a certain sequence of events in the network usually leads to a decrease in network performance, a machine learning model could identify this pattern and alert you to take preventive action.

Another important aspect of network optimization is to ensure efficient data routing. IoT networks can become incredibly complex, with potentially thousands or even millions of devices connected. Efficient data routing ensures that data packets take the most efficient path from the source to the destination. Techniques like multi-path routing can help optimize the data flow, by allowing data packets to take different paths based on network conditions, thus improving the overall network performance.

Remember, an optimized IoT network is not just about speed and performance. It's about ensuring that your IoT devices can reliably communicate with each other and with the cloud, to deliver the functionality that your IoT system is designed for. With the right network management and optimization strategies in place, you can make sure that your IoT system operates smoothly and efficiently, providing you with the valuable data and insights you need.

Module 6 Exam Questions

1. What does IoT stand for?

a) Internet of Technology

b) Internet of Things

c) Interconnected Online Technology

d) Interconnected Online Things

2. Which of the following is a common wireless technology used in IoT?

a) Wi-Fi

b) Bluetooth

c) Zigbee

d) All of the above

3. Which of the following is an example of a Low-Power Wide-Area Network (LPWAN) technology?

a) LTE-M

b) Wi-Fi

c) Bluetooth

d) Ethernet

4. What does the term "network topology" refer to?

a) The physical layout of a network

b) The way data travels in a network

c) Both a and b

d) Neither a nor b

5. What is the main purpose of a gateway in an IoT network?

a) To transfer data from the devices to the cloud

b) To analyze data

c) To provide electricity to the devices

d) To cool down the devices

6. Which of the following could be a challenge when managing IoT networks?

a) Device heterogeneity

b) Scalability

c) Data security

d) All of the above

7. What does interoperability in IoT mean?

a) The ability of different IoT systems and devices to work together

b) The ability of IoT devices to operate continuously without power

c) The ability of IoT devices to interpret data

d) The ability of IoT devices to connect to the internet

8. What is Quality of Service (QoS) in IoT networking?

a) The maximum data transfer rate of a network

b) The percentage of packets that are sent but never reach their destination

c) The performance level of a service or a network

d) The number of devices connected to a network

9. Which of the following could be a part of QoS management in IoT?

a) Traffic prioritization

b) Monitoring network performance

c) Using QoS tools

d) All of the above

10. Which of the following is not an IoT network protocol?

a) HTTP

b) MQTT

c) CoAP

d) JPEG

11. What is the role of an actuator in an IoT system?

a) To collect data

b) To analyze data

c) To perform an action based on data

d) To transmit data

12. What is the main function of edge computing in IoT?

a) To increase data security

b) To reduce latency

c) To provide power to the devices

d) To store data

13. In the context of IoT, what is a "smart" device?

a) A device that can connect to the internet

b) A device that can collect, process, and exchange data

c) A device that can perform tasks without human intervention

d) All of the above

14. What does LPWAN stand for?

a) Large Power Wide Area Network

b) Limited Power Wide Area Network

c) Low Power Wide Area Network

d) Long Power Wide Area Network

15. Which of the following network topologies is typically used in IoT?

a) Star topology

b) Mesh topology

c) Bus topology

d) All of the above

16. Which of the following is a data and message protocol used in IoT?

a) HTTP

b) MQTT

c) CoAP

d) All of the above

17. What parameter does the term "latency" refer to in IoT networking?

a) The maximum data transfer rate of a network

b) The time it takes for a packet of data to travel from source to destination

c) The variability in latency

d) The percentage of packets that are sent but never reach their destination

18. Which of the following sectors can significantly benefit from IoT?

a) Agriculture

b) Healthcare

c) Smart Homes and Buildings

d) All of the above

19. What is one of the main challenges in implementing Quality of Service (QoS) in IoT networking?

a) Device Diversity

b) Network Dynamics

c) Scalability

d) All of the above

20. What is the main purpose of using Low-Power Wide-Area Networks (LPWAN) in IoT?

a) To enable long-range communication

b) To allow for high data rate communication

c) To ensure high power consumption

d) None of the above

Answers:

1. b
2. d
3. a
4. c
5. a
6. d
7. a
8. c
9. d
10. d
11. c
12. b
13. d
14. c
15. d
16. d
17. b
18. d
19. d
20. a

Module 7. IoT Platforms and Ecosystems

Introduction to IoT Platforms and Ecosystems

Welcome to Module 7 on IoT Platforms and Ecosystems. In this module, you will delve deep into the multifaceted world of IoT platforms and ecosystems, crucial elements in the broader IoT landscape.

The Internet of Things is not just about individual devices and sensors, but also about how these elements are seamlessly connected and managed. This is where IoT platforms come in. They form the backbone of any IoT solution, providing the infrastructure needed to ingest, process, and manage data from multiple sources, as well as to integrate various IoT devices and applications.

In this module, we will demystify the concept of an IoT platform and provide you with a clear understanding of its role and functions within an IoT ecosystem. We will uncover the rich variety of capabilities that these platforms offer, from device management and data ingestion to analytics, security, and application enablement.

Not all IoT platforms are created equal, though. As you navigate through this module, we will compare popular IoT platforms in the market, each with their own strengths, weaknesses, and unique selling propositions. We will delve into both commercial and open-source solutions, helping you understand the trade-offs and considerations that should guide your choice of platform for your specific use case.

Beyond platforms, we will explore the broader IoT ecosystem. This includes the various software, hardware, and service elements that come together to form a complete IoT solution. Understanding this ecosystem is essential to grasping the complexities and potential of IoT, as it highlights the interaction and interdependence of various layers and components within an IoT architecture.

As we journey through the module, we will illustrate concepts with practical examples, from industrial automation to smart homes, showing you how IoT platforms and ecosystems truly come to life. By the end of this module, you will not only have a thorough understanding of what an IoT platform is and what it can do, but also how to evaluate and select the right platform for your IoT initiatives.

So, get ready to dive into the world of IoT platforms and ecosystems, where the magic of connectivity, scalability, and interoperability really happens. The journey may be complex, but the rewards - in terms of insights, efficiencies, and opportunities - are immense.

IoT Platform Overview

As we delve deeper into the Internet of Things (IoT) landscape, we arrive at a key component that plays an integral role in the functionality and success of an IoT system: the IoT platform.

An IoT platform is a multi-layer technology that enables straightforward provisioning, management, and automation of connected devices within the IoT ecosystem. In essence, it's the middleware or the support software that connects everything in your IoT system. This includes the devices and sensors, and extends to the applications and services that utilize the data these devices generate.

Now, you might wonder, why do you need an IoT platform? The reason is that it simplifies the process of setting up and managing your IoT system. Let's consider an example. You're building a smart home system. In this system, you have a variety of connected devices - a thermostat, security cameras, lights, and perhaps even a smart refrigerator. Each of these devices generates data that you want to monitor and control, maybe through a smartphone app.

Without an IoT platform, you would need to create separate connections for each device, handle the data individually, and manage different interfaces. This is not only laborious but also impractical as your system scales. An IoT platform, on the other hand, provides a centralized place to manage all these connections, process and store the data, and create applications (like your smartphone app) to interact with the devices.

IoT platforms are multifaceted, providing a variety of capabilities, including:

- **Device management:** This is one of the fundamental roles of an IoT platform. It enables you to connect and manage all your IoT devices, handle firmware and software updates, diagnose and troubleshoot problems, and ensure the devices are working optimally.

- **Data processing and storage:** IoT devices generate a lot of data. Your IoT platform will ingest this data, process it, and store it. It can also provide analytics capabilities to make sense of this data and extract valuable insights.

- **Connectivity and security:** The platform ensures secure and reliable communication between your devices and the cloud, protecting your system from potential security threats.

- **Application enablement:** You want your IoT data to be actionable. This means using it in applications, services, or APIs. Your IoT platform will provide the necessary tools to create and deploy these applications, enabling you to take action based on your IoT data.

- **Integration:** An IoT platform should be able to integrate with other systems, whether it's your existing business software or third-party services. This helps to leverage IoT data across various aspects of your operations.

Now, let's make this more tangible with an example. Let's say you're a farmer looking to implement an IoT system for precision agriculture. You want to deploy a range of sensors across your farm to monitor conditions like soil moisture, temperature, and sunlight intensity. You also want to use this data to automate irrigation and fertilization systems, and you want to track all this information from a single dashboard.

An IoT platform would be the crucial link that makes this possible. It would allow you to connect and manage all your sensors, creating a unified network out of them. It would process and store the data these sensors generate, providing you with real-time insights about your farm's conditions.

You could then use these insights to set up automated systems - say, triggering your irrigation system when the soil moisture drops below a certain level. The platform could even send you alerts when certain conditions are met, like if the temperature rises above a threshold that could be harmful to your crops.

Furthermore, the IoT platform could provide you with tools to create a dashboard, giving you a centralized view of your farm's conditions at any time, from anywhere. And, it would ensure all this data communication happens securely, protecting your system from any potential cyber threats.

That's the power of an IoT platform at work. It simplifies the complexity of IoT, providing you with a toolbox to connect, manage, and build upon your IoT system. However, not all IoT platforms are created equal. They come in different shapes and sizes, each with their own set of features, strengths, and weaknesses. In the following sections, we'll explore some of these platforms, helping you understand how to choose the right one for your specific needs.

Remember, your choice of IoT platform can greatly influence the success of your IoT project. It's the infrastructure upon which your entire IoT system is built. So, take your time to understand it, explore different options, and make an informed decision. Your IoT journey is just beginning.

Comparison of Popular IoT Platforms

In your journey through the Internet of Things (IoT), you will come across a variety of IoT platforms, each offering a unique set of capabilities and strengths. Understanding these differences is crucial to selecting the platform that best suits your specific needs. In this section, we'll examine some of the most popular IoT platforms to help you make an informed choice.

Amazon Web Services (AWS) IoT Core

AWS IoT Core is a managed cloud service that lets connected devices interact with cloud applications and other devices. AWS IoT Core can support billions of devices and trillions of messages, and can process and route those messages to AWS endpoints and other devices reliably and securely.

One of its primary strengths is its tight integration with other AWS services, such as AWS Lambda, Amazon Kinesis, Amazon S3, and Amazon Machine Learning. This provides you with a comprehensive set of capabilities to gather, process, analyze, and act on data generated by connected devices without having to manage any infrastructure.

For example, you could use AWS IoT Core to build an application for a smart city, where various sensors and devices, like traffic lights and air quality sensors, send data to the cloud. Here, the data can be analyzed, and instructions can be sent back to the devices in real-time, helping manage traffic and monitor environmental conditions effectively.

Microsoft Azure IoT Suite

Microsoft's Azure IoT Suite is a comprehensive set of services and solutions that help you move toward the Internet of Things in a straightforward, preconfigured manner. Its key strengths are its compatibility with various operating systems and devices, robust analytics capabilities, and strong security features.

Azure IoT Suite is highly flexible, allowing you to customize solutions according to your needs. It integrates seamlessly with existing Microsoft software, making it a great choice if you're already using Microsoft products.

Suppose you're developing a remote monitoring solution for a wind farm. Using Azure IoT Suite, you can collect real-time data from wind turbines, perform predictive analytics to anticipate maintenance needs, and automate responses to critical alarms, all within a secure environment.

Google Cloud IoT

Google Cloud IoT is a set of fully managed and integrated services that allow you to securely connect, manage, and ingest data from globally dispersed devices. Google Cloud IoT provides a robust application platform with powerful analytics capabilities, making it an excellent choice for data-intensive IoT applications.

With Google Cloud IoT, you can leverage Google's advanced data analytics and machine learning capabilities to gain meaningful insights from your IoT data. For example, if you're building a global supply chain solution, you could use Google Cloud IoT to track goods in real-time, predict potential disruptions using machine learning algorithms, and optimize routes based on real-time traffic data.

IBM Watson IoT

IBM Watson IoT platform offers powerful, pre-integrated tools and services to help you capture and explore data for devices, sensors, and systems. It features advanced analytics, AI capabilities, and blockchain integration, making it a versatile choice for various IoT applications.

Suppose you're working on a smart building solution. With IBM Watson IoT, you could monitor energy usage in real-time, predict equipment failures, and even integrate with blockchain to create transparent and secure transaction records for building occupants.

Bosch IoT Suite

Bosch IoT Suite is a flexible, open standards-based platform for developing, deploying, and managing IoT applications. It supports multiple IoT domains, such as connected cars, smart homes, and industrial IoT.

Bosch IoT Suite is particularly strong in device connectivity and management, making it a good choice if you're dealing with a large number of IoT devices. For example, if you're developing an IoT system for connected vehicles, Bosch IoT Suite could help you manage software updates for the vehicles, monitor their status in real-time, and enable new services like predictive maintenance or usage-based insurance.

ThingWorx IoT Platform

ThingWorx, from PTC, is a complete, end-to-end IoT platform designed for industrial IoT applications. It offers a broad set of features, including device connectivity, machine learning capabilities, augmented reality (AR) integration, and more.

One of the unique strengths of ThingWorx is its easy-to-use interface and rapid application development features. If you're an industrial manufacturer looking to improve operational efficiency, ThingWorx can help you connect and monitor your machines, predict and prevent equipment failures, and even provide AR-powered instructions to help field technicians repair machines faster.

Particle

Particle is a scalable, reliable, and secure Internet of Things device platform that enables businesses to quickly and easily build, connect and manage their connected solutions. It's a great choice for companies looking to develop IoT products, thanks to its integrated hardware, software, and connectivity solutions.

Suppose you're a startup developing a smart home product. Particle can provide you with the hardware modules, the software to control them, and the cellular or Wi-Fi connectivity to get them online. This integrated solution can significantly simplify your development process, allowing you to focus on creating a great user experience.

These platforms are just a few of the many options available to you. Each one offers unique features and capabilities that can benefit different types of IoT projects.

The goal is not to find the "best" IoT platform in a general sense, but to find the one that's best for you - the one that aligns most closely with your needs, capabilities, and objectives.

The best one for you will depend on your specific needs, existing infrastructure, and long-term IoT goals. As you explore these options, consider factors like scalability, integration capabilities, security features, and cost, along with the specific requirements of your IoT project.

Remember, an IoT platform is more than just a piece of technology—it's the foundation of your IoT system. Choosing the right platform can make your IoT journey smoother, more efficient, and more successful. So take the time to explore, ask questions, and carefully evaluate your options. Your perfect IoT platform is out there, waiting to bring your IoT vision to life.

In the next section , we'll take a closer look at open source and commercial IoT solutions, to provide an even more comprehensive view of the IoT platform landscape.

Open Source and Commercial IoT Solutions

In the world of IoT platforms, you have two main categories to choose from: open source and commercial solutions. Each comes with its own set of advantages and considerations. Understanding the differences between these two types of platforms is key to making an informed decision that best fits your needs.

Open Source IoT Solutions

Open source IoT platforms are built on publicly accessible code that you can modify, extend, and distribute. The open-source nature of these platforms fosters a culture of transparency, collaboration, and rapid innovation, with communities of developers continuously improving and building upon the software.

One of the main advantages of open source platforms is their flexibility. You have the freedom to customize the platform to suit your specific needs, which can be a significant advantage if you're working on a unique or complex IoT project. Additionally, open source platforms are often free to use, which can help keep your costs down.

However, while open source platforms offer great flexibility, they may also require more time and technical expertise to set up and manage. You might need to invest in additional resources for customization, integration, and ongoing maintenance. It's also important to consider the vibrancy and responsiveness of the platform's community, as you'll often rely on them for support and updates.

Example: Eclipse IoT

Eclipse IoT is a leading open source IoT platform, with a robust set of services for device management, messaging, and application enablement. Its active developer community continually contributes to a wide range of projects and tools, making it a dynamic and evolving platform.

For instance, if you're building an IoT solution for environmental monitoring, you can use Eclipse IoT to connect and manage your sensors, process and analyze the collected data, and develop applications that respond to environmental conditions in real-time.

Commercial IoT Solutions

On the other hand, commercial IoT platforms are proprietary products developed and maintained by a company. These platforms often come with comprehensive support and service level agreements, which can be a significant advantage if you lack the time or technical resources to manage the platform yourself.

Commercial platforms often offer a more streamlined and user-friendly experience, with intuitive interfaces, pre-built templates, and easy integration with other software. They're designed to get your IoT system up and running quickly, with minimal setup and configuration.

However, commercial platforms can be more expensive than open source alternatives, particularly at scale. They might also be less flexible, as you're limited to the features and capabilities that the platform provides.

Example: Microsoft Azure IoT Suite

Microsoft Azure IoT Suite is a prime example of a commercial IoT platform. It offers a broad range of services, from device connectivity and data storage to advanced analytics and machine learning capabilities. With Azure, you can quickly and easily build, deploy, and scale your IoT applications, all within a secure and reliable environment.

For example, if you're creating a predictive maintenance system for a manufacturing plant, you can use Azure IoT Suite to connect your machines, collect and store their operational data, and develop machine learning models that predict equipment failures before they occur.

Whether you opt for an open source or commercial IoT platform will depend on a variety of factors, including your technical capabilities, budget, project requirements, and long-term IoT strategy. Open source platforms offer unparalleled flexibility and cost-efficiency, but they may require more technical expertise and management effort. Commercial platforms provide a more streamlined and supported experience, but they can be more costly and less customizable.

As you weigh these options, consider your unique needs and circumstances. What are your technical capabilities and resources? What is your budget? What are your project requirements? What is your long-term IoT strategy? By considering these questions, you can make a more informed decision and choose the platform that's right for you. In the next section, we will explore integration and interoperability, critical considerations when adopting an IoT platform.

Integration and Interoperability

Integration and interoperability are key considerations when choosing between open source and commercial IoT solutions. Your IoT platform needs to seamlessly integrate with your existing systems and be compatible with a wide range of devices and technologies.

Open source platforms often have a wide range of integrations, thanks to their large developer communities. For example, the open source platform Node-RED offers a wide range of pre-built nodes that allow you to integrate with various databases, APIs, and other services. You can use Node-RED to build an IoT solution that integrates with your existing customer relationship management (CRM) system, enabling you to automatically update customer records based on IoT data.

Commercial platforms typically offer pre-built integrations with other products from the same vendor, as well as popular third-party services. For instance, AWS IoT Core integrates seamlessly with other AWS services, allowing you to build complex IoT applications that leverage AWS's powerful data storage, processing, and analytics capabilities. You could use AWS IoT Core to build a smart agriculture system that collects data from soil sensors, stores the data in Amazon S3, analyzes the data using Amazon Athena, and visualizes the results using Amazon QuickSight.

Interoperability is another critical factor to consider. With the vast number of IoT devices, protocols, and standards available, it's essential that your IoT platform can communicate effectively with a wide range of technologies. Both open source and commercial IoT platforms usually support a wide range of IoT protocols and standards, but it's always a good idea to verify that the platform supports the specific technologies you plan to use.

In summary, both open source and commercial IoT platforms offer unique strengths. Open source platforms provide unparalleled flexibility and cost-effectiveness, making them a great choice if you have the technical resources to customize and manage the platform yourself. Commercial platforms offer a more streamlined, user-friendly experience and typically come with comprehensive support, making them a good choice if you prefer a more hands-off approach.

Remember, the best IoT platform for you will depend on your specific needs, capabilities, and goals. Understanding the key differences between open source and commercial IoT platforms will allow you to make a more informed decision and select the platform that best aligns with your IoT vision. In the next section, we'll dive deeper into integration and interoperability, examining their role in the IoT ecosystem and their impact on your IoT journey.

Integration and Interoperability

Integration and interoperability are not just desirable attributes but essential components for any successful IoT solution. Let's explore these concepts in detail.

Integration

Integration in the context of IoT refers to the ability of your IoT platform to connect and work seamlessly with various other systems and services. This could include everything from your existing IT infrastructure (like databases and CRM systems) to third-party services (like analytics tools and cloud storage).

For example, let's imagine you're building a smart retail solution to monitor customer behavior in a physical store. Your IoT sensors, such as cameras and motion detectors, collect data about customer movements and interactions. But that's just the start of the story.

The real value comes when you integrate this data with your existing systems. By feeding this IoT data into your CRM system, you can gain a deeper understanding of your customers, personalize their shopping experience, and ultimately drive more sales. Integration with your inventory management system could allow for real-time stock updates based on customer interactions. Furthermore, integration with cloud-based analytics tools could enable you to analyze this data and gain valuable insights into store performance, customer behavior, and more.

Integration is not always straightforward, but many IoT platforms provide tools and features to make this process easier. These may include APIs, SDKs, and pre-built connectors for popular third-party services. The key is to choose an IoT platform that supports the systems and services you currently use or plan to use in the future.

Interoperability

Interoperability, on the other hand, refers to the ability of different devices and systems to work together, even if they're from different manufacturers or use different protocols. In the diverse and fragmented world of IoT, interoperability is a major challenge but also a critical requirement.

Consider a smart home scenario. You might have smart lights from one manufacturer, a smart thermostat from another, and a smart security system from yet another. For your smart home to function effectively, all these devices need to communicate and work together. This is where interoperability comes into play.

Interoperability can be achieved in several ways. One approach is to use open standards and protocols, which are designed to ensure compatibility between different devices and systems. For example, MQTT and CoAP are open protocols commonly used in IoT for device-to-device communication.

Another approach to achieving interoperability is through IoT platforms that provide device management and connectivity services. These platforms can abstract away the complexity of dealing with different devices and protocols, providing a unified interface for managing and communicating with your devices.

For instance, if you're setting up a smart factory, you might have different types of machines from various manufacturers, all using different communication protocols. A good IoT platform can help manage this complexity, allowing you to control, monitor, and gather data from all your machines through a single interface.

Both integration and interoperability are crucial considerations when selecting an IoT platform. You need a platform that can integrate with your existing systems and services, as well as ensure interoperability between a diverse range of devices. But these are not just technical considerations; they also have a significant impact on the effectiveness of your IoT solution and the value you can derive from it.

Remember, the ultimate goal of IoT is not just to connect devices, but to use the data generated by these devices to drive better decisions, improve efficiency, and create new value. By ensuring good integration and interoperability, you can turn this goal into a reality.

User Interface and Experience in IoT Platforms

When you're selecting an IoT platform for your project, there's a factor that can sometimes be overlooked, yet it's one that can significantly impact your productivity and the ease with which you can achieve your goals: the user interface and overall user experience provided by the platform.

User Interface

The user interface (UI) is the point of human-computer interaction and communication in a device, in this case, the IoT platform. It includes a set of commands or menus through which a user communicates with the program. A good user interface should be intuitive and user-friendly, regardless of the user's technical expertise.

For example, consider a scenario where you're using an IoT platform to manage a network of environmental sensors deployed across a city. These sensors are measuring air quality, temperature, humidity, and other environmental factors.

In this case, a well-designed user interface might provide a visual map-based representation of your sensor network, allowing you to see the status of each sensor at a glance. Clicking on a sensor could bring up a dialog showing detailed information about that sensor, including its current readings, battery level, and connectivity status. A sidebar might provide access to additional functions such as setting alerts, viewing historical data, and exporting data for further analysis.

User Experience

While the user interface focuses on the aesthetics and layout of the platform, user experience (UX) is about how the interface works from the user's perspective. It's about how easy it is to use, how intuitive it is, and how efficiently it allows the user to complete tasks.

Continuing with the previous example, a good user experience might include features like drag-and-drop functionality to easily create complex event rules, or the ability to group sensors into categories for easier management. It might also include helpful features like predictive text search to quickly locate specific sensors, or tooltips that provide additional information when you hover over a button or menu item.

Why It Matters

The quality of the UI and UX in an IoT platform has a direct impact on your efficiency and effectiveness. A poorly designed interface can make it difficult to find the functionality you need, leading to wasted time and frustration. Similarly, a poor user experience can make it hard to perform even simple tasks, slowing you down and reducing your productivity.

On the other hand, a well-designed interface and a positive user experience can make you more productive, help you get more value from your IoT data, and reduce the risk of errors.

For example, if you're managing a large network of sensors, a well-designed user interface can help you quickly identify and resolve issues, keeping your network running smoothly. A positive user experience can make it easy to set up complex event rules and alerts, helping you stay informed about important events and trends in your data.

When evaluating the UI and UX of an IoT platform, it's a good idea to take advantage of free trials or demo versions if they're available. This will give you a chance to get hands-on experience with the platform and assess whether its interface and user experience meet your needs.

Remember, while the technical capabilities of an IoT platform are crucial, don't underestimate the importance of a good user interface and user experience. These aspects can significantly influence the success of your IoT projects, and therefore, deserve careful consideration in your decision-making process. In the next section, we will explore another critical aspect of IoT platforms - scalability and reliability.

Scalability and Reliability of IoT Platforms

The concepts of scalability and reliability are critical factors that can influence the performance, longevity, and overall success of your IoT project. Let's explore these concepts and understand their significance in the context of IoT platforms.

Scalability

Scalability refers to the ability of an IoT platform to handle growth in the number of devices and the volume of data that it needs to process. This is not just about the platform's ability to handle your current needs, but also its capacity to meet your future requirements as your project grows and evolves.

For example, imagine you're deploying an IoT solution for a smart city project. Initially, you might start with a few hundred smart lighting systems across several city blocks. However, as the project expands, you might need to manage thousands or even tens of thousands of devices, including not just smart lights, but also traffic sensors, air quality monitors, and more.

In this scenario, scalability is crucial. Your IoT platform needs to be able to handle this increase in devices and data without experiencing performance issues or requiring significant additional resources. This might involve scaling up, by adding more resources to the platform, or scaling out, by distributing the workload across multiple servers or locations.

Reliability

Reliability, on the other hand, is about the platform's ability to function correctly and consistently over time. This involves not just avoiding crashes and errors, but also maintaining a consistent level of performance, even under heavy load or in the face of network issues or hardware failures.

Let's continue with the smart city example. In such a critical application, reliability is of utmost importance. You need the platform to function reliably 24/7, providing real-time data about the city's infrastructure. Any downtime or performance issues can lead to interruptions in the city's services, potentially causing inconvenience or even safety issues for the city's residents.

Reliability can be achieved through various means, such as redundant hardware, fault-tolerant design, and rigorous testing. But it's also about the platform's ability to recover from failures when they do occur, through features like automatic failover, data backups, and recovery tools.

Why It Matters

Scalability and reliability are not just technical considerations; they're also about the future-proofing and resilience of your IoT project.

A scalable IoT platform allows your project to grow and evolve without being held back by technical limitations. It means that you can add more devices, process more data, and implement more complex functionalities as your needs change and grow.

A reliable IoT platform, meanwhile, ensures that your project can operate consistently and effectively over time. It reduces the risk of downtime and performance issues, helping you to maintain a high level of service and avoid the potential costs and reputational damage associated with failures.

When you're evaluating IoT platforms, be sure to consider their scalability and reliability. Look for platforms that can demonstrate their ability to handle large numbers of devices and high volumes of data, and that have built-in features to ensure reliability and resilience. Remember, your choice of IoT platform can have a significant impact on the success of your IoT project, so it's worth taking the time to make an informed decision.

Cost and Pricing Models in IoT Platforms

Understanding the cost implications and pricing models of IoT platforms is crucial. The cost of IoT platforms can vary significantly based on numerous factors such as the number of devices, volume of data, and the specific features you need. Let's delve into these elements to provide you with a clear picture.

Cost Factors

The first aspect to consider is what drives the cost of an IoT platform. There are several key factors:

- **Number of Devices:** Many platforms charge based on the number of devices connected. It's important to estimate not just the number of devices you're starting with, but also how many you might add as your project scales.

- **Volume of Data:** The amount of data generated by your devices also plays a significant role. This can affect costs in terms of data storage, processing, and transfer.

- **Features and Services:** Different platforms offer different sets of features and services. Higher-end platforms may include advanced analytics, machine learning capabilities, enhanced security features, and dedicated support, which would be reflected in the cost.

- **Infrastructure Costs:** Depending on whether the platform is cloud-based, on-premises, or a hybrid, there may be costs for data storage, computing resources, maintenance, and more.

Pricing Models

Once you understand the cost factors, it's essential to familiarize yourself with the common pricing models in the IoT platform market.

- **Pay-per-Device:** In this model, you pay a fixed fee for each device connected to the platform. This is simple and predictable, making it a good fit if your number of devices is relatively stable.
- **Data Usage:** Some platforms charge based on the amount of data your devices generate. This can be a good fit if your data volumes vary significantly, but it can be less predictable than pay-per-device pricing.
- **Subscription-Based:** Many IoT platforms offer subscription plans, which typically include a package of services for a fixed monthly or annual fee. These plans often have different tiers based on the number of devices, data volume, and available features.
- **Freemium:** Some platforms offer a freemium model where basic services are free, but you pay for premium features or additional capacity. This can be a cost-effective way to get started with IoT, but the costs can increase significantly as you scale or require more advanced features.

Why It Matters

Understanding the cost and pricing model of your IoT platform is crucial as it directly impacts your project's budget and return on investment (ROI).

A cost-effective platform for a small pilot project might not be so economical when you scale up to thousands of devices. Conversely, a platform that seems expensive for a small project might become more cost-effective at scale due to economies of scale.

It's also important to factor in the potential indirect costs, like the cost of downtime if the platform isn't reliable, or the cost of additional development work if the platform doesn't have the features you need.

When choosing an IoT platform, take the time to thoroughly understand its pricing model and calculate the total cost of ownership (TCO) over the expected lifetime of your project. Make sure to account for all possible variables, including growth in the number of devices, increases in data volume, and any additional features you might need in the future.

In the next section, we'll delve into the importance of community and support for IoT platforms, another critical factor in your decision-making process.

Community and Support for IoT Platforms

As you work towards building and deploying your IoT solution, the significance of a vibrant community and robust support cannot be overstated. These two elements can significantly impact your experience with an IoT platform and influence the long-term success of your project. Let's understand why.

The Power of Community

Communities around IoT platforms can be an invaluable resource for you. They are places where users — from beginners to experts — share their experiences, knowledge, and solutions to common problems. The benefits of a strong community are multifold:

- **Problem-solving:** You're not the first person to encounter a specific issue, and you won't be the last. Communities can provide quick and practical solutions to problems that have already been encountered and solved by others.

- **Knowledge sharing:** Communities are a treasure trove of shared knowledge. From best practices to innovative uses of the platform, you can learn a lot from the shared experiences of other users.

- **Networking:** Communities provide opportunities to connect with other users, including industry experts and potential partners. You can share ideas, collaborate on projects, or even find new business opportunities.

For example, if you're working with an open-source platform like Eclipse IoT, you'll find a vibrant community of developers who contribute to the platform, share their code, and help each other solve problems. This can save you significant time and effort in troubleshooting and development.

The Importance of Support

While communities are an excellent resource for solving problems and learning from others, there will be times when you need direct, official support from the platform provider. The level and quality of support can vary significantly between different platforms, and this can have a significant impact on your experience.

Here's why robust support is critical:

- **Issue Resolution:** When you encounter technical issues, having access to timely and effective support can mean the difference between a minor hiccup and a major roadblock in your project.
- **Platform Updates:** IoT platforms are dynamic and frequently updated with new features, bug fixes, and security patches. Good support can help you stay on top of these updates and understand how to use them effectively.
- **Training and Onboarding:** Particularly when you're new to a platform, having access to training resources and onboarding support can significantly shorten your learning curve and help you get your project off the ground more quickly.

For instance, if you're using a commercial platform like Microsoft's Azure IoT, you'll have access to comprehensive support resources, including documentation, tutorials, and direct support from Microsoft's technical team. This can provide a safety net as you build and deploy your IoT solution, ensuring that you can get help when you need it.

Making the Right Choice

As you evaluate IoT platforms, don't overlook the importance of community and support. Look for platforms with active, engaged communities and robust support offerings. Remember that the success of your IoT project depends not just on the technical features of the platform, but also on the resources and support available to you as a user.

In the final section of this module, we will explore how to evaluate and select the right IoT platform for your specific needs and circumstances. It's about pulling together all the factors we've discussed and making an informed, strategic decision.

Case Studies of IoT Platform Implementations

In the final section of this module, let's explore some practical examples of IoT platform implementations. These case studies can offer valuable insights into how different types of platforms are used in real-world scenarios, what challenges they help overcome, and the kind of results you can expect.

Case Study 1: Smart Farming with The Things Network

In the world of agriculture, IoT has revolutionized farming practices, leading to the rise of 'smart farming'. Let's consider a case where a farmer decides to implement a smart farming solution using The Things Network, an open-source IoT platform built around the LoRaWAN protocol.

The goal here is to monitor soil conditions and automate irrigation. Using IoT sensors, the farmer is able to measure soil moisture levels in different parts of the farm. This data is then sent to The Things Network platform through LoRaWAN gateways.

On the platform, you can set up an application to process the incoming data, deciding when and where to irrigate based on soil moisture levels. The result? Optimized water usage, healthier crops, and higher yields, all achieved with minimal manual intervention.

Case Study 2: Industrial IoT with AWS IoT Core

Next, let's look at an industrial use case where a manufacturing company wants to implement predictive maintenance to minimize downtime. In this case, they opt for AWS IoT Core due to its robust features and scalability.

The company installs IoT sensors on its machinery to monitor various parameters like temperature, vibration, and pressure. This data is then sent to AWS IoT Core. Using AWS's machine learning services, you can build models that predict potential failures based on the sensor data.

By integrating these predictive models with the company's maintenance systems, you can automate the generation of maintenance tickets when a potential failure is detected. This enables the company to address maintenance issues proactively, reducing downtime and saving costs.

Case Study 3: Smart Home Automation with Home Assistant

Home Assistant is an open-source home automation platform that focuses on privacy and local control. Let's say you decide to implement a smart home solution using Home Assistant to control and automate various devices in your home.

You can integrate various smart devices into Home Assistant - from smart lights and thermostats to security cameras and voice assistants. The platform allows you to control all these devices from a single dashboard, and you can set up automation rules based on time, device status, or even your location.

For instance, you can automate your lights to turn on when the sun sets, your thermostat to adjust based on the time of day, or your security system to arm itself when you leave home. With Home Assistant, you're able to create a smart home that's truly tailored to your needs.

Learning from Case Studies

These case studies illustrate the versatility and potential of IoT platforms across a range of applications. As you read through them, consider how the chosen platform helped address specific challenges and how the implementation was carried out.

Remember, there's no one-size-fits-all solution in the world of IoT. The right platform for your project will depend on your specific requirements, resources, and constraints. As you move forward in your IoT journey, these case studies can serve as a source of inspiration and learning, helping you make informed decisions about your own IoT platform implementation.

Module 7 Exam Questions

1. What is the primary function of an IoT platform?

a) To store data

b) To analyze data

c) To enable communication between IoT devices and applications

d) To run graphics-intensive applications

2. Which IoT platform is built around the LoRaWAN protocol?

a) AWS IoT Core

b) Google IoT Core

c) The Things Network

d) Microsoft Azure IoT

3. What is an example of a device management protocol in IoT?

a) MQTT

b) AMQP

c) LWM2M

d) HTTP

4. Which aspect of an IoT platform helps in dealing with the data traffic generated by millions of devices?

a) User interface

b) Scalability

c) Integration

d) Interoperability

5. Which of these is NOT a consideration while choosing an IoT platform?

a) Interoperability

b) User interface

c) The color of the platform's logo

d) Community and support

6. Which IoT platform is known for its focus on privacy and local control?

a) Home Assistant

b) AWS IoT Core

c) Google IoT Core

d) The Things Network

7. In an IoT ecosystem, what does edge computing enable?

a) Data processing near the source

b) Data storage in the cloud

c) Data visualization on a dashboard

d) Data loss during transmission

8. What is a common use case for IoT in agriculture?

a) Automating irrigation based on soil moisture levels

b) Predicting stock market trends

c) Improving internet connectivity

d) Enhancing virtual reality experiences

9. What is the primary benefit of using open-source IoT platforms?

a) They are always free

b) They allow customization and offer control over your data

c) They provide 24/7 customer support

d) They come pre-loaded with data

10. Which IoT platform is commonly used for industrial applications like predictive maintenance?

a) AWS IoT Core

b) The Things Network

c) Home Assistant

d) IBM Watson IoT

11. Which of the following is a common pricing model for IoT platforms?

a) Pay-per-use

b) Subscription-based

c) Freemium

d) All of the above

12. What do IoT sensors in smart farming typically monitor?

a) Soil moisture levels

b) Stock prices

c) Internet speed

d) Human emotions

13. Which of these is NOT an IoT communication technology?

a) Bluetooth

b) Wi-Fi

c) LoRaWAN

d) HTML

14. How does the support from the platform provider assist IoT platform users?

a) Issue resolution

b) Updates on platform changes

c) Training and onboarding

d) All of the above

15. What do IoT protocols facilitate?

a) Communication between IoT devices and applications

b) Power supply to IoT devices

c) Physical connection between IoT devices

d) Manufacturing of IoT devices

16. Which of the following is a benefit of a vibrant community around an IoT platform?

a) Problem-solving

b) Knowledge sharing

c) Networking

d) All of the above

17. What is a major security challenge in IoT?

a) Interoperability

b) User interface design

c) Authentication and encryption

d) Subscription pricing

18. In the context of IoT platforms, what does 'interoperability' refer to?

a) The ability to process large amounts of data

b) The ability to work with different devices, applications, and protocols

c) The ability to operate without internet connectivity

d) The ability to work in various geographical locations

19. How does a good User Interface (UI) contribute to an IoT platform?

a) It improves device connectivity

b) It ensures data security

c) It enhances user experience and ease of use

d) It reduces the cost of the platform

20. What is a key advantage of implementing IoT in the agriculture industry?

a) It can help automate irrigation based on soil conditions

b) It can increase internet speed in rural areas

c) It can improve graphics on computer monitors

d) It can help in predictive stock trading

Answers:

1. c
2. c
3. c
4. b
5. c
6. a
7. a
8. a
9. b
10. a
11. d
12. a
13. d
14. d
15. a
16. d
17. c
18. b
19. c
20. a

Module 8. IoT in Industry Verticals

Introduction to IoT in Industry Verticals

Welcome to Module 8 on IoT in Industry Verticals. In this module, you will delve into the remarkable applications of the Internet of Things across a range of industry sectors. We'll explore how these advanced technologies are driving transformation, enhancing operational efficiency, and creating new business opportunities across the globe.

The Internet of Things, often simply referred to as IoT, is not a single technology but rather a complex, intertwined network of systems, devices, and applications. Its potential is vast and diverse, with the capability to impact virtually every industry in significant ways. As we continue our journey through this ever-evolving landscape, we will delve into the specific applications of IoT in various industry verticals.

We'll begin by investigating how IoT technologies are redefining homes and buildings, making them more energy-efficient, safer, and more comfortable. We'll also explore how cities are leveraging IoT to become "smart," optimizing infrastructure, public services, and quality of life for their inhabitants.

In the realm of manufacturing and industry, we'll examine how IoT is fueling the fourth industrial revolution, known as Industry 4.0, by enhancing automation, improving production efficiency, and facilitating predictive maintenance.

In healthcare, we'll discuss how IoT is revolutionizing patient care, monitoring, and diagnostics. From wearable devices to remote patient monitoring, IoT is enabling healthcare providers to deliver personalized, on-demand medical services.

From there, we'll delve into the role of IoT in agriculture, transportation and logistics, energy and utilities, retail, and education. We'll see how it's helping farmers optimize crop yields, enabling more efficient supply chains, transforming energy management, redefining the shopping experience, and enhancing the learning process.

Through the exploration of practical examples and real-world case studies, you will gain a robust understanding of how IoT is being harnessed in different sectors. This knowledge will not only provide you with a solid foundation for understanding the present landscape of IoT but also equip you with the insight to anticipate future trends and opportunities.

In a world where digital and physical realities are increasingly intertwined, the exploration of IoT applications in various industries is not just relevant—it's crucial. Whether you're a student, a technology enthusiast, a professional looking to stay ahead of the curve, or a leader aiming to drive digital transformation in your organization, this module will provide valuable insights and inspirations.

So, get ready to embark on an exciting journey that will take you to the heart of digital transformation across industries, powered by the Internet of Things.

Smart Homes and Buildings

In this section, we will take a closer look at how the Internet of Things (IoT) is revolutionizing the way you live and interact with your physical environment.

The concept of a "smart home" refers to a residence that uses internet-connected devices to enable the remote monitoring and management of appliances and systems. This includes your heating and cooling systems, lighting, security systems, and even your refrigerator, washing machine, and other appliances. It's about making your life more comfortable, convenient, secure, and energy-efficient.

Let's start with the heart of smart home technology: home automation hubs. These devices act as the central point of control for all your smart devices. For example, think of Amazon's Alexa or Google Home. With these devices, you can use voice commands or your smartphone to control various aspects of your home, from adjusting the thermostat and turning off lights to locking doors and even brewing your morning coffee.

Now, let's think about energy efficiency. IoT-enabled thermostats, like the Nest Learning Thermostat, adapt to your lifestyle and adjust the temperature based on your routine. They 'learn' your habits over time. For example, if you typically lower the temperature when you leave for work in the morning, the thermostat will start doing this for you automatically. Not only does this make your life more convenient, but it can also significantly reduce your energy consumption and utility bills.

Safety and security is another critical area where IoT is making a big difference. Think of smart security systems and cameras, like those from Ring or Arlo. You can remotely monitor your home in real-time, receive alerts if any unusual activity is detected, and even communicate directly with whoever is at your door, all from your smartphone.

Smart lighting systems, like those offered by Philips Hue, allow you to control the lighting in your home remotely or set it on a schedule. You can even change the color and brightness of the light to match your mood or the time of day. Imagine your lights gradually brightening in the morning to help you wake up naturally, or changing to a warm, relaxing glow in the evening.

Now, let's extend these concepts to larger buildings. The same principles and technologies can be applied to commercial buildings to create 'smart buildings.' For example, building automation systems can control lighting, HVAC (heating, ventilation, and air conditioning), and security systems to optimize comfort, safety, and energy efficiency. IoT sensors can monitor everything from temperature and humidity to occupancy and light levels, providing valuable data that can be used to improve the building's performance and the comfort of its occupants.

For example, consider a smart office building that uses IoT sensors to monitor occupancy levels in real time. If a particular floor or area is empty, the building's system can automatically turn off the lights and lower the heating or cooling in that area to save energy. Similarly, smart elevators can predict demand based on usage patterns and position themselves accordingly, improving efficiency and reducing wait times.

The potential of smart homes and buildings is vast, and we're only scratching the surface here. As you delve deeper into the world of IoT, you'll discover an array of innovative devices and applications that can make your home and other buildings more connected, efficient, and responsive. The future of living and working is smart, and with the knowledge you're gaining, you're well on your way to becoming a part of it. Welcome to the age of smart homes and buildings.

Smart Cities and Infrastructure

Let's delve into how the Internet of Things (IoT) is revolutionizing urban environments, making them more efficient, sustainable, and enjoyable to live in.

A "smart city" is an urban area that uses different types of electronic methods and sensors to collect data. This data is then used to manage assets, resources, and services efficiently. The goal is to improve the quality of life for you and your fellow citizens by using technology to improve the efficiency of services and meet residents' needs.

Let's start with one of the most fundamental aspects of city living: transportation. In a smart city, IoT devices can monitor traffic flow in real time, allowing city officials to optimize traffic light sequences, reducing congestion and making your commute smoother. In some cities, IoT-enabled smart parking solutions are being deployed. These systems guide you directly to available parking spaces, saving you time and reducing the amount of traffic caused by drivers looking for parking.

Public transportation can also benefit from IoT technologies. Real-time tracking of buses and trains can provide you with accurate arrival times, making your journey more predictable and less stressful. In some cities, this information is integrated into a single app, giving you a comprehensive view of all your transportation options.

Now, let's consider utilities. Smart grids use IoT sensors to optimize the delivery of electricity, reducing energy consumption and costs. They can also identify and respond to outages more quickly, reducing downtime. For you, this means a more reliable and possibly cheaper energy supply. Additionally, IoT devices can help monitor and manage water supply networks, reducing waste and ensuring a more sustainable use of this vital resource.

Waste management is another area where smart cities are using IoT to improve efficiency. Smart waste bins can monitor the level of waste and notify the relevant authorities when they need to be emptied. This can optimize collection routes, reducing operational costs and environmental impact.

Safety and security are also enhanced in smart cities. IoT-enabled surveillance systems can monitor public spaces for unusual activity, improving the safety of your community. Smart street lighting can adjust to the presence of pedestrians, cyclists, or cars, improving visibility and safety while saving energy.

In terms of urban infrastructure, smart buildings, like the ones we discussed in the previous section, are an essential component of smart cities. But beyond individual buildings, IoT can help manage and optimize entire districts, integrating energy management, transportation, and other services to create a more sustainable and enjoyable living environment.

To visualize this, imagine living in a district where the heating in your home is powered by a district heating network that uses waste heat from local industries. Your electric car is charged at a nearby charging station that's powered by solar panels. The lighting in the public spaces responds to the time of day and presence of people, and you're guided to available parking spaces by a real-time parking management system.

That's the vision of a smart city. It's a place where technology is used to improve the efficiency of services and make your life more comfortable and sustainable. As we continue our journey into the world of IoT, we'll see that this vision is already becoming a reality in cities around the world. And as you deepen your understanding of these technologies, you'll be better equipped to play a role in shaping this future. Welcome to the era of smart cities and infrastructure.

Industrial IoT (IIoT) and Manufacturing

Welcome to a world where machines communicate with each other, processes are optimized in real-time, and decision-making is enhanced with unparalleled data insights. This is Industrial IoT (IIoT) in action within the manufacturing sector, a technological revolution that will redefine your approach to production, maintenance, and even design.

IIoT is the application of IoT technologies in industrial settings, often with an emphasis on automation, integration of machines, and real-time data exchange. IIoT is not simply about connecting devices; it's about building new workflows, improving efficiency, and unlocking new value from your existing systems.

Consider the manufacturing floor: once characterized by manual operations, it is now increasingly automated and data-driven. With IIoT, you can connect every machine, sensor, and device in your factory to a single network. This allows machines to communicate with each other and with you, leading to unprecedented levels of automation and visibility into your operations.

For example, imagine a conveyor system in your factory. With IIoT, sensors on the conveyor belts can monitor the speed, temperature, and load. If a belt becomes overloaded or a motor overheats, the system can automatically slow down or shut off to prevent damage. You can even program it to send you an alert, allowing you to intervene promptly, if necessary. This not only reduces the risk of equipment damage but also enhances your operational efficiency and productivity.

Another key aspect of IIoT in manufacturing is predictive maintenance. Traditionally, you might have followed a schedule for maintenance or waited until a machine broke down. But with IIoT, sensors can monitor machine performance and detect signs of potential problems before they cause a failure. For instance, a sensor might detect abnormal vibrations or temperatures that suggest a bearing is wearing out. The system can then alert you, so you can replace the bearing during a planned downtime, avoiding an unexpected breakdown that could disrupt your production schedule.

Moreover, IIoT can provide you with valuable insights into your manufacturing processes. By analyzing data from your machines, you can identify bottlenecks, inefficiencies, or quality issues. For instance, you might discover that a particular machine is slower than others, causing a backlog, or that a certain process has a high rate of errors. With these insights, you can make informed decisions about where to invest in upgrades, adjustments, or training to improve your overall efficiency and product quality.

IIoT can also facilitate the implementation of digital twins - virtual replicas of your physical assets, processes, and systems. With digital twins, you can simulate and analyze different scenarios, enabling you to optimize your processes, predict outcomes, and innovate faster. For example, if you're planning to introduce a new product line, you can use a digital twin to simulate the production process, identify potential issues, and fine-tune your operations before you even start the actual production.

Remember, the power of IIoT comes not just from the technology itself, but how you use it to transform your operations. Understanding your needs and applying IIoT strategically will allow you to drive significant improvements in efficiency, quality, and flexibility in your manufacturing operations. And as you embark on this journey, remember that IIoT is not a destination, but a continuous journey of improvement and innovation.

Healthcare and IoT

In this point of our journey, we will explore how the Internet of Things (IoT) is transforming healthcare, making it more patient-centric, efficient, and personalized.

IoT's integration into healthcare, often termed as the Internet of Medical Things (IoMT), opens up a whole new world of possibilities. At its core, it's about connecting devices, applications, and systems to create a network that can monitor, collect, exchange, analyze, and deliver valuable insights to improve patient outcomes and streamline healthcare delivery.

Let's start with the most common application of IoT in healthcare: wearable devices. From fitness trackers like Fitbit to smartwatches like the Apple Watch, these devices can continuously monitor your vital signs such as heart rate, sleep patterns, activity levels, and more. They provide a wealth of data that not only helps you understand and manage your health but can also be shared with your healthcare provider to facilitate more personalized care.

For patients with chronic conditions such as diabetes or heart disease, IoT offers even more profound benefits. For instance, smart glucose meters and insulin pumps can help manage diabetes by continuously monitoring blood sugar levels and delivering precise insulin doses. Similarly, IoT-enabled heart rate monitors can alert you and your healthcare provider to potentially dangerous irregularities in your heart rhythm.

Moving onto telemedicine, IoT can enable remote patient monitoring, a method particularly useful for elderly patients or those living in remote areas. With IoT-enabled devices, you can have your health parameters like blood pressure, oxygen levels, or weight monitored from the comfort of your home. The data is automatically sent to your healthcare provider, who can then make informed decisions about your care without the need for you to make a trip to the clinic.

Now, let's consider how IoT can improve medication management. Smart pillboxes, for example, can remind you when it's time to take your medication and even alert your healthcare provider if you miss a dose. This can be particularly useful if you have a complex medication schedule.

In the context of hospitals, IoT can streamline operations and improve patient care in numerous ways. IoT-enabled asset tracking, for example, can ensure that essential equipment like wheelchairs or defibrillators are always available when needed. In patient rooms, IoT devices can monitor environmental conditions such as temperature and humidity to ensure optimal comfort and prevent the spread of infections.

One of the most revolutionary applications of IoT in healthcare is in the field of predictive analytics. By analyzing data from a multitude of IoT devices, healthcare providers can identify patterns and trends that can predict health issues before they become serious. For instance, subtle changes in your heart rate, activity levels, or sleep patterns could signal the onset of a health issue, potentially allowing for preventive intervention.

In conclusion, IoT is revolutionizing healthcare by shifting the focus from reactive to preventive care, making healthcare more personalized, efficient, and accessible. As you navigate through the world of IoT, you will increasingly come across technologies that not only enhance healthcare delivery but also empower you to take control of your health. Welcome to the era of Healthcare and IoT.

Agriculture and IoT

Now, we will explore how the Internet of Things (IoT) is transforming the agricultural sector, paving the way for precision farming and creating opportunities for more sustainable and efficient food production.

The use of IoT in agriculture, often referred to as Smart Farming or Precision Agriculture, leverages connected devices to monitor and manage various agricultural processes, thereby optimizing resources, improving yields, and reducing environmental impact.

Let's start with one of the most fundamental applications of IoT in agriculture: soil monitoring. IoT-enabled soil sensors can continuously monitor parameters like soil moisture, temperature, and nutrient levels. This real-time data allows you to make informed decisions about irrigation and fertilizer application, ensuring that your crops get exactly what they need when they need it. The result? Improved crop health, higher yields, and reduced resource wastage.

Next, consider the role of IoT in irrigation. Smart irrigation systems use data from soil sensors, weather forecasts, and other sources to optimize irrigation schedules and volumes. These systems can automatically adjust watering based on the specific needs of your crops, reducing water consumption while ensuring optimal growth conditions.

Now, let's discuss pest and disease management. IoT devices equipped with cameras or other sensors can monitor your crops for signs of pests or disease. By identifying potential issues early, you can intervene promptly, minimizing crop damage and reducing the need for chemical treatments.

Livestock farming is another area where IoT is making a significant impact. IoT-enabled devices such as GPS collars and health monitoring sensors can track the location, health, and behavior of your livestock in real time. This not only enhances animal welfare by enabling early detection of health issues but also improves farm management. For instance, tracking the grazing patterns of your livestock can help manage pasture more effectively.

In the realm of farm machinery, IoT is facilitating the rise of smart and autonomous equipment. Tractors and other machinery equipped with GPS and IoT technology can perform tasks like plowing, planting, and harvesting more accurately and efficiently. For example, a smart tractor could adjust its speed and the amount of seed it disperses based on the soil's fertility, ensuring uniform crop growth.

One of the most promising aspects of IoT in agriculture is its potential for data analysis and decision making. By collecting data from various IoT devices, you can gain a comprehensive understanding of your farm's operation. Advanced analytics can identify trends, predict yields, and provide insights to guide your decisions, from crop selection and planting schedules to resource management and market timing.

In conclusion, IoT is revolutionizing agriculture by enabling precision farming, which is about doing the right thing, in the right place, at the right time, in the right way. As you deepen your understanding of IoT technologies, you'll be better equipped to harness these tools to make your farming operations more sustainable, efficient, and profitable. Welcome to the era of Agriculture and IoT.

Transportation and Logistics

In this section we will explore how the Internet of Things (IoT) is transforming the way goods and people move around the world, leading to improved efficiency, safety, and sustainability.

Let's start by looking at how IoT is revolutionizing fleet management. IoT devices installed in vehicles can monitor a wide range of parameters, including location, speed, fuel consumption, and engine performance. This real-time data can help you manage your fleet more efficiently. For instance, by analyzing fuel consumption and driving patterns, you can identify ways to reduce fuel costs and emissions. You can also schedule maintenance based on actual vehicle usage, which can help prevent breakdowns and prolong the lifespan of your vehicles.

Next, let's consider the role of IoT in logistics and supply chain management. IoT sensors can track the location and condition of goods as they move through the supply chain. For example, if you're shipping perishable goods, IoT sensors can monitor temperature and humidity levels during transit, alerting you if conditions deviate from the optimal range. This can help prevent spoilage, reduce waste, and ensure the quality of your goods when they reach their destination.

In the context of warehousing, IoT can help optimize inventory management. IoT-enabled smart shelves can monitor inventory levels in real-time, automatically alerting you when it's time to restock. This can reduce the risk of stockouts or overstocking, improving inventory turnover and reducing storage costs.

In terms of last-mile delivery, IoT can help improve efficiency and customer satisfaction. For example, IoT-enabled route optimization can determine the most efficient delivery route considering current traffic conditions, reducing delivery times and fuel consumption. Real-time tracking allows you and your customers to know exactly where the delivery vehicle is and when the delivery will arrive.

Now, let's consider public transportation. IoT can improve the efficiency and user experience of public transit systems. For example, IoT sensors can monitor vehicle locations and passenger counts in real time, enabling dynamic route optimization based on current demand. This can reduce wait times and crowding, improving the user experience. At the same time, real-time data can help transit operators manage their fleet more efficiently.

Safety is another critical area where IoT can make a significant impact in transportation. For instance, IoT devices can monitor driver behavior, vehicle conditions, and environmental factors to identify potential safety risks. This could range from alerting you if a driver is showing signs of fatigue, to warning you if a vehicle part is showing signs of wear and tear.

In conclusion, IoT is transforming transportation and logistics by providing real-time visibility and insights into every aspect of the movement of goods and people. As you navigate through the world of IoT, you'll discover how these technologies can help you improve efficiency, reduce costs, enhance safety, and make transportation and logistics more sustainable. Welcome to the era of Transportation and Logistics with IoT.

Energy and Utilities

In this point of our exploration, we'll delve into how the Internet of Things (IoT) is reshaping the energy sector, enabling smarter and more sustainable use of resources.

One of the most transformative applications of IoT in this sector is the development of smart grids. A smart grid uses a network of IoT devices to monitor and manage the flow of electricity in real-time. For example, if a power line fails, the smart grid can instantly detect the problem and reroute electricity to minimize downtime. It can also balance electricity supply and demand more efficiently, reducing the need for energy storage and lowering costs.

Connected to this idea is the concept of smart meters. These IoT devices can monitor your energy consumption in real-time and communicate this information back to the energy provider. This not only allows for more accurate billing based on actual usage but also enables dynamic pricing, where energy prices can be adjusted based on supply and demand. This can encourage you to use energy more efficiently, such as by running high-energy appliances during off-peak times when energy is cheaper.

Another exciting area is the integration of renewable energy sources into the grid. IoT can help manage the variability of renewables by monitoring weather conditions and predicting energy production. For example, IoT devices can forecast how much electricity your solar panels will produce based on the current weather forecast, allowing the grid to adjust accordingly.

Now, let's shift our focus to the home environment. IoT can help you manage your energy consumption more efficiently through smart home devices. For example, a smart thermostat can learn your routines and adjust the heating and cooling of your home to save energy while maintaining comfort. Similarly, smart appliances can operate during off-peak energy times to reduce costs and strain on the grid.

In the context of water utilities, IoT can help detect leaks and monitor water quality in real-time. For instance, IoT sensors in your home's water system can alert you if they detect unusual water usage, which could indicate a leak. At a larger scale, IoT devices can monitor the quality of water in reservoirs or treatment plants, alerting utility providers to any issues that need attention.

Moving onto the oil and gas industry, IoT devices can monitor pipeline infrastructure for leaks or faults, improving safety and reducing environmental impact. They can also optimize extraction operations by providing real-time data on drilling conditions and equipment performance.

In conclusion, IoT is transforming the energy and utilities sector by improving efficiency, sustainability, and service reliability. As you deepen your understanding of IoT technologies, you'll be better equipped to make informed decisions about your energy usage and contribute to a more sustainable future. Welcome to the era of Energy and Utilities with IoT.

Retail and IoT

In this section, we will explore how the Internet of Things (IoT) is revolutionizing the retail industry, enhancing the shopping experience, and enabling more efficient operations.

IoT is transforming retail through smart stores. IoT-enabled devices like smart shelves and digital signage are creating interactive, personalized shopping experiences. Smart shelves can track inventory in real-time, alerting you when it's time to restock. Digital signage can display targeted advertisements based on shopper demographics or even specific products they are looking at or have in their shopping cart.

One of the most impressive applications of IoT in retail is the smart fitting room. Here, interactive mirrors can suggest outfits based on the items you bring into the room, show you different color options, or even let you request different sizes without leaving the fitting room. This not only enhances your shopping experience but also provides valuable insights to retailers about your preferences and shopping habits.

Now, let's consider the role of IoT in supply chain management. IoT devices can track products from the manufacturing plant to the store shelf, providing real-time visibility into the supply chain. This can help you manage inventory more efficiently, reduce stockouts and overstocks, and ensure product authenticity.

Another critical aspect of IoT in retail is customer insights. IoT devices can collect a wealth of data about shopping behavior, from the paths customers take through your store to the products they spend the most time looking at. Analyzing this data can provide valuable insights to help you optimize store layout, merchandising, and marketing strategies.

IoT is also making strides in enhancing security in retail. For instance, IoT-enabled video surveillance systems can monitor your store in real-time and use advanced analytics to detect suspicious behavior. Similarly, smart locks and alarm systems can provide better control over who has access to different areas of your store, improving security and reducing the risk of theft.

Beacons are another fascinating IoT application in retail. These small, wireless devices communicate with shoppers' smartphones as they move around your store, sending personalized offers or information based on their location. This can enhance the shopping experience, increase customer engagement, and drive sales.

In the context of e-commerce, IoT can help improve logistics and delivery. For example, IoT devices can monitor warehouse inventory levels, automate order picking, and track shipments in real-time, improving efficiency and customer satisfaction.

In conclusion, IoT is transforming the retail industry by creating smarter, more connected shopping experiences, improving operational efficiency, and providing valuable customer insights. As you explore the world of Retail and IoT, you'll discover how these technologies can help you create a more successful and customer-centric retail business. Welcome to the era of Retail with IoT.

Education and IoT

Now we will examine how the Internet of Things (IoT) is playing a transformative role in the education sector, creating innovative learning environments, and enhancing administrative efficiency.

IoT is revolutionizing the classroom experience. In a smart classroom, IoT devices such as interactive whiteboards and tablets can facilitate collaborative, interactive learning experiences. For instance, these tools can enable you to annotate a shared document in real time or manipulate 3D models of complex concepts, enhancing understanding and engagement.

Another exciting application of IoT in education is personalized learning. Using data from IoT devices, learning platforms can adapt to your learning style, pace, and interests. For example, if you're struggling with a particular concept, the platform can suggest additional resources or change the way the material is presented to help you understand better.

Let's now turn our attention to the role of IoT in campus safety. IoT devices can help create a safer learning environment through real-time video surveillance, smart locks, and emergency notification systems. For example, if an unauthorized person attempts to access your classroom or dormitory, an IoT-enabled security system can send you an alert.

In the context of administrative efficiency, IoT can help streamline operations and reduce costs. For instance, IoT sensors can monitor energy usage across campus, identifying opportunities to save energy and reduce carbon footprint. Similarly, smart waste management systems can optimize waste collection routes and schedules based on real-time data, improving efficiency and sustainability.

Another area where IoT can make a significant difference is in the management of physical assets. IoT devices can track the location and condition of equipment like laptops, projectors, and laboratory instruments, helping to prevent loss or damage and ensure these resources are available when you need them.

On a broader scale, IoT can help improve campus navigation and transportation. For instance, IoT devices can provide real-time information about bus schedules or the availability of parking spaces, making it easier for you to get around campus.

In terms of remote learning, IoT can create more engaging and effective online learning experiences. For instance, IoT devices can monitor your engagement and provide feedback to help you stay focused and motivated. In some cases, IoT devices can even create immersive virtual or augmented reality experiences, allowing you to explore virtual labs or historical sites from your living room.

In conclusion, IoT is transforming education by creating more engaging, personalized learning experiences, enhancing safety, and improving administrative efficiency. As you delve deeper into the world of Education and IoT, you'll discover how these technologies can enhance your learning journey and help you succeed in the digital age. Welcome to the era of Education with IoT.

IoT in Entertainment and Media

Let's explore how the Internet of Things (IoT) is revolutionizing the entertainment and media industry, creating immersive experiences and enabling personalized content delivery.

Starting with your home entertainment system, IoT is creating a connected entertainment experience. Smart TVs, speakers, and gaming consoles can communicate with each other to provide a seamless entertainment experience. For example, you can control your entire home entertainment system through a single device, making it easier to find and enjoy your favorite content.

Another fascinating application of IoT in entertainment is the rise of smart venues. In a smart stadium or concert hall, IoT devices can enhance your experience in several ways. For instance, you can use your smartphone to find your seat, order food or merchandise, and even participate in interactive experiences during the event.

IoT is also making waves in the gaming industry. With IoT, gaming devices can collect data about your play style and preferences, enabling developers to create more engaging and personalized gaming experiences. In addition, IoT can facilitate more immersive gaming experiences through virtual reality (VR) and augmented reality (AR) technologies.

In the realm of media, IoT is enabling more personalized content delivery. For instance, your smart TV or streaming device can analyze your viewing habits to recommend content tailored to your interests. This not only enhances your viewing experience but also helps media companies to retain and engage their audience more effectively.

Let's also consider the role of IoT in content creation. IoT devices like drones and 360-degree cameras are enabling new forms of storytelling, from immersive VR experiences to stunning aerial footage. For instance, as a content creator, you can use a drone to capture unique perspectives that were previously inaccessible or too expensive to achieve.

In the area of advertising, IoT is enabling more targeted and interactive ad experiences. For example, digital billboards can display ads based on real-time data, such as the time of day, weather conditions, or even the demographics of the people nearby. Similarly, interactive kiosks can engage you in interactive ad experiences, such as virtual product try-ons.

On a broader scale, IoT can help media companies gain valuable insights into audience behavior. For instance, IoT devices can track which programs are being watched, when, and for how long, providing data that can inform programming and advertising decisions.

In conclusion, IoT is transforming the entertainment and media industry by creating more immersive, personalized, and interactive experiences. As you explore the world of Entertainment and Media with IoT, you'll discover how these technologies can enhance your entertainment experiences and provide new opportunities for creativity and engagement. Welcome to the era of Entertainment and Media with IoT.

Module 8 Exam Questions

1. What is an example of an IoT device used in smart homes?

a) Smart TV

b) Smart fridge

c) All of the above

d) Smart thermostat

2. Which technology is used in retail to send personalized offers to shoppers' smartphones?

a) NFC

b) Beacons

c) RFID

d) Bluetooth

3. How does IoT contribute to enhancing the classroom experience?

a) By personalized learning

b) Through interactive whiteboards

c) Both a and b

d) By providing real-time video surveillance

4. Which sector benefits from IoT through real-time inventory tracking and order picking?

a) Education

b) Agriculture

c) E-commerce

d) Healthcare

5. In the context of smart cities, what does IoT enable?

a) Traffic management

b) Waste management

c) All of the above

d) Water management

6. How is IoT utilized in healthcare?

a) Telemedicine

b) Remote patient monitoring

c) Predictive care

d) All of the above

7. Which IoT device can be used in agriculture for soil analysis?

a) Smart irrigation systems

b) Drones

c) GPS trackers

d) Soil sensors

8. In the transport and logistics sector, what is an application of IoT?

a) Predictive maintenance

b) Fleet management

c) All of the above

d) Real-time tracking

9. How does IoT contribute to energy and utilities management?

a) Energy consumption monitoring

b) Smart grid systems

c) Predictive maintenance

d) Both a and b

10. What IoT technology is used to provide immersive experiences in the gaming industry?

a) Drones

b) Virtual reality (VR)

c) Both a and b

d) Augmented reality (AR)

11. Which sector uses IoT devices for real-time surveillance and safety?

a) Healthcare

b) Education

c) All of the above

d) Retail

12. How does IoT enhance the shopping experience in retail?

a) Digital signage

b) Smart fitting rooms

c) All of the above

d) Smart shelves

13. How does IoT contribute to personalized learning in education?

a) By suggesting additional resources

b) By adapting to your learning style

c) By changing the way material is presented

d) All of the above

14. What is the role of IoT in supply chain management in retail?

a) Inventory management

b) Real-time product tracking

c) Ensuring product authenticity

d) All of the above

15. In smart cities, what does IoT contribute to?

a) Smart parking

b) Air quality monitoring

c) Traffic control

d) All of the above

16. How is IoT used in agriculture for crop management?

a) Climate monitoring

b) Soil analysis

c) All of the above

d) Precision farming

17. In entertainment and media, how does IoT enhance your home entertainment system?

a) It allows interactive gaming experiences

b) It enables control of the entire system through a single device

c) It recommends content tailored to your interests

d) All of the above

18. How is IoT contributing to security in the retail industry?

a) Smart locks

b) IoT-enabled video surveillance

c) Alarm systems

d) All of the above

19. What is an application of IoT in the energy and utilities sector?

a) Grid management

b) Smart metering

c) Leak detection

d) All of the above

20. How does IoT aid in creating a safer learning environment in education?

a) Real-time video surveillance

b) Smart locks

c) Emergency notifications systems

d) All of the above

Answers:

1. c
2. b
3. c
4. c
5. c
6. d
7. d
8. c
9. d
10. b
11. c
12. c
13. d
14. d
15. d
16. c
17. d
18. d
19. d
20. d

Module 9. IoT Trends and Future Outlook

Introduction to IoT Trends and Future Outlook

Welcome to Module 9 on IoT Trends and Future Outlook. In this module, you will embark on a fascinating journey into the rapidly evolving landscape of the Internet of Things (IoT) as we examine current trends, anticipate future developments, and explore the potential impact these changes will have on our lives and society at large.

The world of IoT is dynamic, and its pace of evolution is unrelenting. The advancements we've seen so far are just the beginning. The promise of billions of connected devices communicating and interacting with each other in real-time paints a vivid picture of a future where physical and digital realms seamlessly intertwine.

In this module, we will take a closer look at the cutting-edge technologies emerging in the IoT space. From enhanced connectivity options and improved data analytics to sophisticated AI integration and more sustainable, power-efficient devices, we will delve into the innovations shaping the future of IoT.

We'll also explore the IoT market and its adoption trends. The IoT ecosystem is increasingly finding resonance across industries, with use cases spanning from smart homes to smart cities, healthcare, agriculture, and beyond. Understanding these trends can provide insights into future opportunities and challenges that professionals, businesses, and consumers may encounter.

The future of IoT is not without its challenges. As the technology continues to mature, we must grapple with a host of issues, from security and privacy concerns to technical hurdles, and ethical dilemmas. This module will give you a comprehensive understanding of these challenges and discuss potential solutions and approaches to address them.

Furthermore, we will probe the ethical considerations and the social impact of IoT. The widespread adoption of IoT has profound implications for society. We will scrutinize these implications and discuss the need for ethical guidelines and regulations in the development and deployment of IoT technologies.

The future of IoT is a topic of immense interest and importance, not just for tech enthusiasts, but for anyone who is part of this increasingly interconnected world. Whether you're a student, a professional, an entrepreneur, or simply a curious mind, this module aims to prepare you for the IoT-driven future, arming you with the knowledge and insights to navigate this exciting terrain confidently.

Join us as we delve into the intriguing world of IoT, deciphering the trends of today to anticipate the realities of tomorrow. Welcome to the future. Welcome to Module 9: IoT Trends and Future Outlook.

Emerging IoT Technologies

In the world of the Internet of Things (IoT), evolution is constant. As we delve into emerging IoT technologies, it's important to keep in mind that these advancements are not only shaping our present but are the building blocks of the future. They are the very technologies that are going to augment your capabilities, streamline your operations, and transform your world.

Let's first talk about 5G and Beyond. The advent of 5G represents a quantum leap in connectivity. With its high speed, low latency, and the ability to connect a colossal number of devices simultaneously, 5G is set to revolutionize how your IoT devices communicate. Imagine downloading a high-definition movie in seconds or operating your remote devices with near-zero delay. That's the potential of 5G. But the journey doesn't stop at 5G; research into 6G has already begun, promising even greater enhancements.

Next up are Edge Computing and Fog Computing. With the avalanche of data produced by IoT devices, transferring all data to a cloud for processing is becoming increasingly impractical. This is where Edge Computing comes into play. It allows data to be processed closer to where it's generated - on your devices themselves, reducing latency, and improving response times. Similarly, Fog Computing extends the Cloud closer to your devices, providing a distributed infrastructure that can manage the storage, computation, and network resources more efficiently.

Artificial Intelligence (AI) and Machine Learning (ML) are two transformative technologies that are increasingly becoming integral to IoT. They bring the power of learning and decision making to your IoT devices. For instance, your smart home system could learn your daily patterns and adjust the heating, lighting, and even your coffee machine to suit your routine. Or consider predictive maintenance, where AI can learn from machine behavior to predict and alert you about potential breakdowns before they occur.

Digital Twins represent another fascinating frontier in IoT. A digital twin is a virtual replica of a physical entity. This could be anything from a product, a system, a process, or even a person. For instance, you could have a digital twin of your manufacturing process, allowing you to simulate and test different scenarios without disrupting your actual operations. You can optimize your processes, predict outcomes, and make informed decisions based on the insights gained.

Blockchain technology, known for its decentralized and secure nature, is making its way into IoT as well. You can use blockchain to create a secure and tamper-proof ledger of transactions between your IoT devices, enhancing the security and trust in the data they generate and share.

Another exciting area of development is in Energy Harvesting technologies. These technologies can draw power from the environment - be it light, heat, or motion - to fuel your IoT devices. This could dramatically extend the lifespan of your devices and reduce the need for battery replacements.

Last but not least, advancements are also being made in IoT Hardware. From more powerful and compact sensors to low-energy and flexible electronics, these innovations are enabling the creation of IoT devices that are smaller, cheaper, more energy-efficient, and adaptable to a wider range of environments.

Exploring these emerging technologies opens up a realm of possibilities. They offer new ways to leverage the power of IoT in your life and work. They bring you closer to a future where you're not just connected but are a part of a smart, responsive, and efficient world. As we move forward, the potential of these technologies will only grow, offering you an exciting palette of opportunities to innovate, optimize, and transform.

IoT Market and Adoption Trends

As the Internet of Things (IoT) continues to evolve, so too does its market and the trends that drive its adoption. Understanding these trends is critical as they influence your choice of IoT technologies, strategies, and implementation models. So, let's dive into the current market and adoption trends in IoT.

One of the most significant trends in the IoT market is the accelerated growth of connected devices. It's estimated that by 2025, there will be over 75 billion IoT devices worldwide. This exponential growth is transforming every facet of your life, from your home to your office, and even your city.

Your home is increasingly becoming a hub of smart devices, with everything from your thermostat to your refrigerator, your security system to your lighting, becoming interconnected. This trend, known as the Smart Home, is driven by your desire for convenience, energy efficiency, and enhanced security.

In the realm of work, the Industrial Internet of Things (IIoT) is revolutionizing manufacturing, logistics, and other industrial sectors. IIoT allows you to monitor and control your production lines, optimize your supply chain, and ensure the safety of your workforce. For instance, you could use sensors to track the condition of your equipment and predict maintenance needs, minimizing downtime and saving costs.

The rise of Smart Cities is another major trend to watch. From traffic management systems that adjust to real-time conditions, to smart grids that optimize energy consumption, to waste management systems that notify when bins need emptying, these technologies are making cities more efficient, sustainable, and livable.

On a broader scale, the adoption of IoT in healthcare is gaining momentum. Wearable devices that monitor your health parameters, smart pills that deliver medication precisely, remote patient monitoring systems – these innovations are improving patient care and outcomes. They're helping you manage your health proactively and giving healthcare providers a deeper understanding of your needs.

The agricultural sector is also experiencing a transformation with the advent of IoT. Precision farming techniques powered by IoT devices are enabling you to monitor soil conditions, track livestock, automate irrigation, and much more. These advancements are increasing crop yields, improving animal welfare, and promoting sustainable farming practices.

In the retail sector, IoT is enhancing both the online and in-store shopping experience. With IoT, you can provide personalized recommendations, streamline the checkout process, manage your inventory efficiently, and gain deep insights into your customer behavior.

The transportation and logistics industry is another sector where IoT is making a significant impact. From fleet management to smart logistics solutions, IoT is helping streamline operations and improve efficiency. For instance, with IoT, you can track your fleet in real-time, optimize routes based on traffic and weather conditions, monitor driver behavior, and predict maintenance needs.

One of the most promising trends is the increased integration of Artificial Intelligence (AI) and Machine Learning (ML) with IoT. These technologies are enabling your IoT devices to not only collect data but learn from it, make predictions, and take decisions. This integration is opening up a whole new realm of possibilities, from smart homes that adjust to your lifestyle, to predictive maintenance in factories, to personalized shopping experiences in retail.

Another notable trend is the growing focus on IoT security. As the number of connected devices grows, so do the potential security risks. To counter these risks, there's a growing emphasis on developing secure IoT devices and systems, implementing robust security protocols, and educating users about safe IoT practices.

Lastly, the advancement in connectivity technologies like 5G, Low-Power Wide-Area Networks (LPWAN), and Wi-Fi 6 is fueling the growth of IoT. These technologies are providing the high-speed, low-latency connections that are essential for the effective functioning of sophisticated IoT systems. For instance, 5G, with its potential to connect a massive number of devices and offer ultra-reliable low-latency communication, is set to transform industries by enabling real-time remote control, automation, and advanced AI applications. The impact of this can be seen in areas such as autonomous vehicles, smart cities, and telemedicine, where instantaneous data transfer and real-time responses are crucial.

LPWAN technologies, on the other hand, are facilitating the development of IoT applications that require wide area coverage and low power consumption. This is particularly relevant for applications like smart agriculture, environmental monitoring, and smart metering, where IoT devices are dispersed over a wide area and are often battery-powered.

Wi-Fi 6, with its improved capacity, performance, and efficiency over previous Wi-Fi standards, is expected to play a significant role in the deployment of IoT in high-density environments like smart homes, enterprise networks, and urban areas. Its ability to handle more devices simultaneously and offer more data throughput makes it well-suited for IoT applications that involve a high number of connected devices or high data transfer rates.

In conclusion, advancements in these connectivity technologies are not just enhancing the capabilities of IoT systems, but also expanding the scope of potential IoT applications, paving the way for the future growth and evolution of IoT.

Future Challenges for IoT

As the Internet of Things (IoT) continues to expand and evolve, it brings with it a host of challenges that you will need to address. These challenges stem from various aspects of IoT, including security, privacy, technology, and more. In this section, we will delve into the most pressing future challenges for IoT and discuss some practical examples.

One of the most critical challenges you will face in the IoT landscape is security. With billions of connected devices, the potential attack surface for hackers and cybercriminals is enormous. Cyberattacks on your IoT devices could lead to the theft of sensitive data, disruption of critical infrastructure, and even physical harm. To counter this threat, you must invest in robust security measures, including encryption, authentication, and regular firmware updates, while staying vigilant about emerging threats and vulnerabilities.

Closely related to security is the challenge of data privacy. IoT devices collect vast amounts of data, some of which may be highly sensitive, such as your health data, personal information, or financial transactions. Ensuring that this data is securely stored and transmitted, and that your privacy is respected, is crucial. You need to adopt privacy-by-design principles and adhere to relevant regulations, such as GDPR, to ensure that your IoT solutions protect your users' privacy.

Another significant challenge is the issue of interoperability and standardization. With a multitude of IoT devices, platforms, and protocols available, ensuring seamless communication and interaction between them can be daunting. You must adopt open standards and invest in technologies that facilitate interoperability to ensure that your IoT ecosystem functions smoothly and can integrate with future innovations.

The rapid growth of IoT devices also raises concerns about sustainability and environmental impact. Many IoT devices rely on batteries, which can be resource-intensive to produce and challenging to dispose of responsibly. To address this challenge, you need to explore energy-efficient technologies, such as energy harvesting, and design devices that have a longer lifespan and are made from eco-friendly materials.

Scalability is another challenge that you will need to tackle as the IoT ecosystem expands. As the number of connected devices grows, your IoT infrastructure must be capable of handling the increased load, from data storage and processing to communication and device management. Embracing technologies like edge computing and fog computing can help you achieve the necessary scalability while maintaining efficiency and performance.

Bandwidth and network congestion are also potential challenges in the future of IoT. The sheer volume of data generated by IoT devices can strain network resources, leading to slower response times and degraded performance. To mitigate this issue, you need to invest in advanced connectivity technologies, such as 5G and LPWAN, and employ techniques like data compression and adaptive bitrate streaming to optimize bandwidth usage.

Another challenge in the IoT domain is talent acquisition and workforce development. As IoT technologies continue to evolve and become more specialized, there is an increasing demand for skilled professionals who can design, develop, and maintain IoT solutions. To address this challenge, you must invest in education and training programs, partner with universities and research institutions, and nurture a culture of continuous learning and innovation.

Lastly, the rapid growth of IoT brings with it ethical and social challenges. As IoT technologies permeate every aspect of your life, questions arise about the potential impact on your privacy, autonomy, and even your mental well-being. As a stakeholder in the IoT ecosystem, you must consider these ethical implications and work towards developing guidelines and regulations that balance the benefits of IoT with the need to protect your values and rights.

The challenges outlined above are by no means exhaustive, but they represent some of the most pressing issues that you will need to address as the IoT landscape continues to evolve. Understanding these challenges and taking proactive steps to mitigate them will allow you to ensure that your IoT solutions are secure, reliable, sustainable, and respectful of user rights.

As you continue to deploy more and more IoT devices, the challenge of device lifecycle management becomes increasingly critical. You'll need to manage device firmware updates, handle device failures and replacements, and retire devices securely when they reach the end of their lifecycle. This requires robust device management solutions that can help you maintain the health and security of your IoT devices throughout their lifecycle.

Legal and regulatory compliance is another challenge that is gaining prominence. As IoT technologies become more pervasive, they are attracting greater scrutiny from regulatory bodies. You will need to keep abreast of evolving laws and regulations related to IoT, such as those concerning data protection, cybersecurity, and telecommunications, and ensure that your IoT solutions are compliant.

In the realm of Artificial Intelligence (AI) and Machine Learning (ML), which are increasingly being integrated with IoT, there are unique challenges as well. These include the explainability and transparency of AI/ML decisions, potential biases in AI/ML algorithms, and the need for robust validation and testing of AI/ML models. Addressing these challenges requires a strong understanding of AI/ML principles, rigorous testing methodologies, and ethical guidelines for AI use.

Lastly, as the IoT ecosystem becomes more complex, the challenge of system complexity and manageability also grows. You will need to manage an intricate web of devices, networks, platforms, and services, each with their own quirks and requirements. This calls for sophisticated management tools that can provide visibility into your IoT ecosystem, automate routine tasks, and help you troubleshoot issues efficiently.

The future of IoT is ripe with opportunities, but it also presents numerous challenges. As you continue your journey in the IoT world, keep these challenges in mind. Equip yourself with the knowledge, strategies, and tools to tackle them head-on. Remember, every challenge is an opportunity for innovation and improvement. And with the right approach, you can turn these challenges into stepping stones towards a more connected, intelligent, and sustainable future.

Ethical Considerations and Social Impact

As you navigate the Internet of Things (IoT) landscape, it is critical to consider the ethical implications and social impact of this technology. Like any powerful tool, IoT can be used for good or ill, and understanding these potential impacts can guide you in using it responsibly and effectively.

One of the primary ethical considerations in IoT is privacy. IoT devices are often always-on, collecting data about your daily activities and behaviors. This constant surveillance has the potential to erode your personal privacy, and it's crucial to balance the benefits of IoT with the need for privacy. For instance, while a smart home security system can provide safety and peace of mind, it also collects data about your comings and goings, which if mishandled or misused, could intrude on your privacy. You should ensure that any IoT solution you use or design respects privacy, includes clear user consent, and follows best practices for data handling and storage.

Another ethical concern is data security. With the vast amount of sensitive data collected by IoT devices, there is a risk of that data being stolen or misused. A breach of your IoT device could lead to identity theft, financial loss, or other harmful outcomes. Therefore, it's vital to prioritize the security of your IoT devices, implementing robust encryption, secure authentication, and regular security updates.

The issue of ownership and control of data is another ethical consideration. As you use IoT devices, you generate a wealth of data. But who owns this data, and who has the right to control and use it? These questions are often not clearly answered, leading to potential ethical dilemmas. You should be aware of these issues and advocate for clear, fair, and transparent data ownership policies.

A further consideration is the potential for bias and discrimination. As IoT devices and systems increasingly use Artificial Intelligence (AI) and Machine Learning (ML) algorithms, there's a risk that biases in the data or in the design of these algorithms could lead to unfair outcomes. For instance, an AI-powered hiring tool might discriminate against certain groups if it's trained on biased data. You must strive to understand and mitigate these biases, ensuring that your IoT solutions are fair and equitable.

On a societal level, the impact of IoT is profound. The widespread use of IoT has the potential to transform communities and social structures. For instance, smart cities can improve the quality of life by optimizing traffic, reducing pollution, and providing better public services. However, they can also lead to increased surveillance and potential inequities if not implemented thoughtfully.

Furthermore, the rise of IoT might lead to significant shifts in the job market. While IoT can automate routine tasks and improve efficiency, it can also displace jobs, especially in sectors like manufacturing and transportation. At the same time, it can create new jobs in areas like IoT development, data science, and cybersecurity. You must be aware of these shifts and work towards a future where the benefits of IoT are shared equitably.

The adoption of IoT in healthcare is another area where significant social impact is expected. While IoT can improve patient care and empower individuals to manage their health, it also raises questions about data privacy, access to healthcare services, and the potential for health disparities. For example, while a wearable health tracker can help you manage your health, it also collects sensitive health data that must be protected. Additionally, access to such technology may be limited in underserved communities, potentially widening health disparities.

Lastly, the environmental impact of IoT is an important consideration. While IoT has the potential to improve energy efficiency and promote sustainable practices, the production, usage, and disposal of billions of IoT devices also carry a significant environmental footprint. Therefore, you should strive to incorporate sustainability principles in your IoT projects.

This could involve selecting energy-efficient devices, adopting designs that minimize waste and resource usage, and implementing end-of-life management strategies for IoT devices, such as recycling or repurposing. It's also essential to consider renewable energy sources for powering IoT devices, especially in large-scale deployments.

Another aspect to consider is the reduction of electronic waste (e-waste). With the rapid pace of IoT development, devices can quickly become obsolete, leading to a significant amount of e-waste. Strategies such as modular design, where components can be upgraded rather than replacing the whole device, can help mitigate this issue.

Furthermore, using IoT technologies to monitor and reduce environmental impact across various sectors, such as smart grids for energy management, precision agriculture for efficient use of water and fertilizers, and smart buildings for optimizing energy usage, can significantly contribute to sustainability goals.

In essence, as you navigate the future of IoT, it's crucial to balance technological advancements with environmental responsibility, ensuring a sustainable and ethical path forward in this exciting field.

The Future of IoT

As you continue to explore the world of the Internet of Things (IoT), it's essential to not only understand the present landscape but also envision what the future may hold. The future of IoT is poised to be exciting, transformative, and full of opportunities and challenges. Let's delve into what this future might look like.

One of the most anticipated advancements in the future of IoT is the widespread implementation of 5G technology. With its high data speed, low latency, and ability to connect a vast number of devices, 5G will significantly enhance the capabilities of IoT networks. Imagine downloading a full-length high-definition movie in seconds on your smartphone or experiencing near real-time communication between devices in a smart city. The advent of 5G is set to make such scenarios a reality.

Next, the future of IoT will see a deeper integration of Artificial Intelligence (AI) and Machine Learning (ML). The vast amount of data generated by IoT devices presents a rich resource for AI and ML algorithms. These technologies can help make sense of this data, providing valuable insights, predicting trends, and automating decision-making processes. Think of a smart home system that learns your daily routine and adjusts the lighting, temperature, and even your coffee machine accordingly, creating a personalized and efficient living environment.

Edge computing is another trend that will shape the future of IoT. By processing data on the device itself or a local server, rather than sending it to a remote data center, edge computing reduces latency, conserves bandwidth, and enhances privacy. This could be particularly beneficial in scenarios where quick data processing is critical, such as autonomous vehicles, where a split-second delay in data processing can make the difference between a safe journey and a collision.

The future will also see a rise in autonomous things, from self-driving cars to drones to robotic process automation. IoT will play a crucial role in enabling these technologies by providing connectivity, real-time data exchange, and remote monitoring and control. Imagine a world where your car drives you to work, drones deliver your online shopping, and robots handle tedious administrative tasks, freeing up your time for more important and enjoyable activities.

IoT in healthcare is also set to transform the way you manage your health and interact with healthcare services. Wearable devices will monitor your health in real-time, telemedicine will allow you to consult with healthcare professionals remotely, and smart hospitals will use IoT to improve patient care and streamline operations. This could mean early detection of health issues, more convenient access to healthcare services, and a more personalized and efficient healthcare experience.

The concept of Digital Twins, virtual replicas of physical systems, is gaining traction and is set to become a crucial tool in the IoT arsenal. These digital representations allow you to simulate, predict, and optimize the functioning of IoT systems, leading to improved efficiency and reduced costs. For instance, a digital twin of a wind turbine can help you optimize its performance, anticipate maintenance needs, and minimize downtime.

In the industrial sector, the rise of Industrial IoT (IIoT) will revolutionize manufacturing, logistics, and other industries. IIoT combines IoT technology, AI, and other innovations to automate industrial processes, improve productivity, and enhance safety. This might mean predictive maintenance of machinery, real-time supply chain visibility, and safer, more efficient work environments.

The growing focus on sustainability will also influence the future of IoT. As the environmental impact of technology becomes a pressing concern, there will be an increasing demand for IoT solutions that promote energy efficiency, reduce waste, and contribute to a circular economy. This could involve smart grids that optimize energy use, IoT-enabled waste management systems, or smart buildings that minimize their environmental footprint.

While the future of IoT is full of potential , it's important to remember that these advancements will come with their own set of challenges. The increased complexity of IoT systems, concerns about data privacy and security, the ethical implications of AI and automation, and the environmental impact of IoT devices are all issues that you will need to navigate.

However, these challenges also present opportunities for innovation, learning, and growth. By staying informed about the latest developments, understanding the implications of new technologies, and adopting a proactive and responsible approach, you can help shape a future where IoT brings significant benefits to individuals, societies, and the planet.

In the future, the distinction between the physical and digital worlds may continue to blur. Your home, workplace, city, and even your body could be interconnected in an intricate web of IoT devices, creating a world where information flows seamlessly and digital intelligence enhances every aspect of your life. This might seem like science fiction, but with the rapid advancements in IoT, this future could be closer than you think.

In conclusion, the future of IoT is promising and holds endless possibilities. As you venture further into the world of IoT, remember that you are not just a passive observer of this future, but an active participant. Your actions, decisions, and innovations can help shape the future of IoT. So embrace the opportunities, tackle the challenges, and look forward to a future where IoT transforms the way you live, work, and interact with the world.

Module 9 Exam Questions

1. What is one of the anticipated advancements in the future of IoT?

a) 2G technology

b) 3G technology

c) 4G technology

d) 5G technology

Which of these technologies will play a significant role in the future of IoT due to the vast amount of data generated by IoT devices?

a) Quantum Computing

b) Blockchain

c) Artificial Intelligence (AI) and Machine Learning (ML)

d) Holography

What is edge computing in the context of IoT?

a) Processing data on a central server

b) Processing data on the device itself or a local server

c) Processing data in the cloud

d) Processing data manually

In what area is IoT expected to make significant impacts in the future by monitoring health in real-time and providing remote healthcare services?

a) Sports

b) Education

c) Retail

d) Healthcare

What is a digital twin in the context of IoT?

a) A backup copy of digital data

b) A virtual replica of a physical system

c) A duplicate IoT device

d) A clone of a digital AI

Which technology combines IoT, AI, and other innovations to automate industrial processes?

a) Residential IoT

b) Commercial IoT

c) Industrial IoT (IIoT)

d) Educational IoT

Which of the following is not a primary ethical consideration in IoT?

a) Privacy

b) Data security

c) Control of data

d) Energy efficiency

What is the main advantage of 5G technology in the context of IoT?

a) Lower cost

b) High data speed and low latency

c) Better design

d) Longer device lifespan

Which of these is an example of an autonomous thing enabled by IoT?

a) Smartphones

b) Self-driving cars

c) Desktop computers

d) E-books

What could the IoT device of a wind turbine help with in the context of a digital twin?

a) Optimize its performance

b) Anticipate maintenance needs

c) Minimize downtime

d) All of the above

What could a smart home system that integrates AI and IoT do?

a) Learn your daily routine

b) Adjust the lighting and temperature according to your routine

c) Control your coffee machine according to your routine

d) All of the above

Which of these is a risk associated with the data collected by IoT devices?

a) Identity theft

b) Financial loss

c) Both a and b

d) None of the above

What is a potential ethical dilemma associated with the data generated by IoT devices?

a) Data ownership

b) Data storage

c) Data speed

d) Data appearance

What is one way that IoT can impact social structures?

a) By transforming communities

b) By increasing surveillance

c) By improving public services

d) All of the above

Which sector could be particularly affected by job displacement due to IoT?

a) Manufacturing

b) Transportation

c) Both a and b

d) None of the above

How can IoT contribute to sustainability?

a) By promoting energy efficiency

b) By reducing waste

c) By contributing to a circular economy

d) All of the above

How can the concept of Digital Twins be beneficial in IoT?

a) By simulating IoT systems

b) By predicting the functioning of IoT systems

c) By optimizing the functioning of IoT systems

d) All of the above

What aspect of IoT is likely to be enhanced with the widespread implementation of 5G technology?

a) Data speed

b) Latency

c) Connectivity of a vast number of devices

d) All of the above

In what field is the IoT expected to bring significant transformations by enabling real-time health monitoring and telemedicine?

a) Entertainment

b) Agriculture

c) Healthcare

d) Tourism

What is one of the challenges that the increased complexity of IoT systems might present?

a) Data privacy concerns

b) Security concerns

c) Ethical implications of automation

d) All of the above

Answers:

1. d
2. c
3. b
4. d
5. b
6. c
7. d
8. b
9. b
10. d
11. d
12. c
13. a
14. d
15. c
16. d
17. d
18. d
19. c
20. d

Module 10. Building a Career in IoT

Introduction to Building a Career in IoT

Welcome to Module 10 on Building a Career in IoT. This is an exciting and crucial part of your IoT learning journey where we transition from knowledge acquisition to practical application. You will learn about the diverse range of opportunities available in the fast-growing field of Internet of Things (IoT) and how to navigate your way to a successful career in this domain.

In this module, we will demystify the various job roles and responsibilities associated with IoT, unraveling the skills and competencies that are most in demand. We'll provide a roadmap to help you identify and pursue the right educational pathways and certifications to bolster your IoT career prospects.

But it's not just about what you know, it's also about who you know. Therefore, we'll delve into the importance of networking and professional associations in IoT. We'll explore the top industry sectors and employers utilizing IoT technology and look at current trends in the IoT job market to provide you with the most current and relevant career information.

With the IoT landscape evolving at a fast pace, understanding how to future-proof your career is essential. We'll share tips for a successful IoT career, highlighting the importance of continuous learning and adaptability. If you have entrepreneurial aspirations, we'll also discuss the exciting world of IoT startups.

Through real-world career stories and interviews, you'll get firsthand insights into the experiences of those who have walked the path before you. Finally, we'll gaze into the future of IoT jobs and skills, helping you to stay ahead of the curve and prepare for what's next.

So, are you ready to take the next step in your IoT career journey?

Job Roles and Opportunities in IoT

In this fast-paced digital world, IoT has become a significant game-changer, leading to an unprecedented demand for professionals skilled in this domain. As you begin your journey into the realm of IoT, you'll discover a variety of job roles and opportunities that can suit different interests and skill sets. Let's delve into some of these exciting roles.

IoT Developer

As an IoT developer, your primary responsibility would be to create and implement IoT solutions. This role involves designing and writing software for embedded systems and IoT devices, working with data scientists to develop algorithms, and managing data integration, connectivity, and system requirements. You may also be involved in testing, debugging, and improving IoT applications to ensure their efficiency and functionality. If you have a strong background in programming languages such as Python, Java, or C, and an understanding of networking and system engineering, this role could be an excellent fit for you.

IoT Solution Architect

If you like the idea of overseeing the overall strategy and execution of IoT projects, the role of an IoT Solution Architect might be right for you. This role requires a deep understanding of IoT technology and the ability to design comprehensive IoT solutions that meet business needs. As an IoT Solution Architect, you'll work closely with various teams, including software developers, data scientists, and business leaders, to ensure the successful implementation of IoT systems. To succeed in this role, you will need a strong technical background, strategic thinking, and excellent communication skills.

IoT Systems Engineer

IoT Systems Engineers are responsible for the overall engineering and management of IoT systems. This includes designing, implementing, and maintaining the hardware, software, and networking systems required for IoT solutions. As an IoT Systems Engineer, your role would involve ensuring the seamless integration of different IoT components and resolving any technical issues that arise. If you enjoy problem-solving and have a strong background in systems engineering, this role could be a great choice.

IoT Security Specialist

With the exponential growth in IoT devices, security has become a major concern. As an IoT Security Specialist, you would be responsible for identifying potential security risks and implementing measures to mitigate them. This role requires a deep understanding of cybersecurity principles and IoT technologies. If you have a passion for security and a knack for identifying potential threats, this role could be a great fit for you.

IoT Data Analyst

The operation of IoT devices generates a vast amount of data. This is where the role of an IoT Data Analyst comes in. Your job would be to analyze this data and provide insights that can help improve IoT systems and drive business decisions. If you have strong analytical skills, a background in statistics or data science, and an interest in IoT, this role might be just what you're looking for.

These are just a few of the many roles in the IoT field. Other opportunities include IoT Project Manager, IoT Network Specialist, IoT Product Manager, and more. Remember, as the IoT landscape continues to evolve, new roles and opportunities are likely to emerge.

Of course, IoT isn't just about the technical roles. There are also numerous management and leadership roles that are critical to the successful implementation and operation of IoT systems. Let's explore a few of these.

IoT Project Manager

As an IoT Project Manager, you will be responsible for overseeing and managing the execution of IoT projects from inception to completion. This includes planning, coordinating with different teams, managing resources, and ensuring that projects are completed on time and within budget. If you have strong organizational and leadership skills and an understanding of IoT technologies, this role can be an ideal choice.

IoT Product Manager

If you have an interest in the strategic aspect of IoT, the role of an IoT Product Manager might be right for you. In this role, you'll oversee the development and management of IoT products, ensuring they meet customer needs and achieve business goals. You'll work closely with various teams, including development, marketing, and sales, and will play a critical role in defining the product vision, strategy, and roadmap. Strong communication, strategic thinking, and customer empathy are key skills for this role.

IoT Operations Manager

As an IoT Operations Manager, your job would be to ensure the smooth operation of IoT systems within an organization. This includes managing the technical team, overseeing system maintenance and upgrades, and resolving any operational issues. You'll also be responsible for ensuring the security and compliance of IoT systems. If you have a strong technical background, excellent leadership skills, and an understanding of operational management, this role could be a great fit for you.

IoT Business Development Manager

This role is for those who have a knack for spotting business opportunities and building relationships. As an IoT Business

Development Manager, you'll be responsible for identifying new business opportunities, building and maintaining relationships with clients, and promoting the organization's IoT solutions. This role requires a good understanding of IoT technology, strong communication skills, and a strategic mindset.

Chief IoT Officer (CIoTO)

At the executive level, the role of the Chief IoT Officer is emerging in some organizations. The CIoTO is responsible for defining the strategic direction of IoT initiatives within the organization, ensuring alignment with overall business objectives. This role involves a mix of technical understanding, strategic planning, and leadership.

Remember, whether you're more interested in hands-on technical work or prefer leading and managing teams, there's a place for you in the IoT industry. By understanding your interests and strengths, you can choose the right path and start building the skills needed for a successful IoT career.

The key to building a successful career in IoT is to understand your interests and strengths, and align them with the right role. So, take some time to reflect on what you've learned about these roles. Consider which ones resonate with you the most, and then start building the skills and competencies needed for your chosen path. The world of IoT is vast and full of opportunities, and it's waiting for you to dive in!

Required Skills and Competencies

Now that you have an understanding of the diverse job roles in IoT, it's time to focus on the specific skills and competencies required to excel in these roles. As with any career path, the exact skills you'll need will vary depending on the specific job role you're interested in. However, there are several key skills and competencies that are commonly required across most IoT roles.

Technical Skills

Firstly, it's important to have a solid foundation of technical skills relevant to IoT. These include:

- **Programming:** Proficiency in programming languages such as Python, Java, C, and JavaScript is often required in IoT roles. These languages are commonly used for developing IoT applications and systems.

- **Networking:** Understanding networking concepts such as TCP/IP, WiFi, LTE, and Bluetooth is crucial for IoT roles, as IoT devices often rely on these technologies to communicate.

- **Hardware Interfacing:** Knowledge of hardware interfacing and sensors is essential, as IoT involves a lot of interaction between hardware and software.

- **Cybersecurity:** With the increasing connectivity of devices comes increased security risks. Therefore, understanding cybersecurity principles and how to apply them to IoT is critical.

- **Data Management and Analytics:** IoT devices generate vast amounts of data. Being able to manage and analyze this data, using tools such as SQL or Hadoop and techniques such as machine learning, is a valuable skill in the IoT field.

Project Management Skills

Whether you're in a leadership role or not, having project management skills can greatly benefit your career in IoT. These include:

- **Planning and Organization:** Being able to plan and organize tasks effectively is crucial in ensuring IoT projects are completed on time and within budget.

- **Risk Management:** Understanding how to identify and mitigate potential risks in IoT projects can help prevent costly and time-consuming issues.

- **Communication and Team Coordination:** IoT projects often involve teams from different disciplines. Being able to communicate effectively and coordinate teams is key to the successful implementation of IoT projects.

Soft Skills

While the technical and project management skills are important, don't underestimate the value of soft skills. These include:

- **Problem-Solving:** IoT involves a lot of problem-solving, whether it's figuring out how to integrate different devices or troubleshooting technical issues.

- **Critical Thinking:** Being able to analyze situations critically and make informed decisions is a valuable skill in IoT, where you'll often need to evaluate different solutions and choose the most effective one.

- **Adaptability:** The IoT field is constantly evolving. Being adaptable and willing to continuously learn and update your skills is crucial to staying relevant in the IoT industry.

- **Collaboration:** IoT projects often involve working with people from different disciplines. Being able to collaborate effectively is important in ensuring the success of IoT projects.

Keep in mind that these are just some of the many skills and competencies required in IoT roles. Depending on the specific role you're interested in, you may need to develop additional skills. However, these core skills provide a solid foundation for a career in IoT. As you progress in your IoT journey, focus on building and strengthening these skills, and you'll be well-prepared for the exciting opportunities that lie ahead in the IoT field.

IoT Industry Sectors and Employers

The Internet of Things (IoT) is a transformative technology that's impacting a wide range of industry sectors. As a result, there are numerous potential employers you could work for, ranging from tech startups to multinational corporations. Let's delve deeper into some of the key sectors where IoT is making a significant impact and the potential employers within these sectors.

Technology and IT Services

This sector includes companies that provide hardware, software, and services related to IoT. These could be big tech giants, innovative startups, or specialized IoT companies.

For example, you could work for a company like Microsoft or Amazon Web Services, both of which offer IoT platforms and services. Alternatively, you might find opportunities at a company like ARM or Intel, which design and manufacture the microchips used in many IoT devices.

Telecommunications

Telecommunication companies play a crucial role in IoT, providing the networks that enable IoT devices to communicate. This includes traditional telecom companies, as well as companies focused on specific IoT connectivity technologies like LoRaWAN or NB-IoT.

For instance, you could work for telecom companies like Verizon or AT&T, both of which are heavily involved in the IoT space. Or, you might find opportunities at a company like Sigfox, which specializes in IoT connectivity.

Manufacturing and Industrial

The manufacturing and industrial sector is a major adopter of IoT technologies, using it for applications such as predictive maintenance, automation, and supply chain optimization. This opens up opportunities at a wide range of manufacturing and industrial companies.

Companies like Siemens, General Electric, and Bosch, which are implementing IoT in their industrial operations, could be potential employers. Or, you might find opportunities at a company like Rockwell Automation, which provides IoT solutions specifically for the manufacturing sector.

Healthcare

The healthcare sector is leveraging IoT to improve patient care, enhance efficiency, and drive innovation. This includes hospitals, health tech companies, and biotech firms.

Companies like Philips and Medtronic, which develop IoT-enabled medical devices, could offer intriguing roles. Or, you could find opportunities at a health tech startup developing innovative IoT solutions for healthcare.

Transportation and Logistics

IoT is transforming transportation and logistics, enabling real-time tracking, route optimization, and predictive maintenance. This includes companies in shipping, logistics, and automotive industries.

Companies like DHL or UPS, which are using IoT to optimize their logistics operations, could be potential employers. Similarly, automotive companies like Tesla or BMW, which are integrating IoT technologies into their vehicles, offer exciting opportunities.

Remember, these are just a few examples of the many sectors and employers in the IoT industry. With the widespread adoption of IoT, opportunities are emerging in virtually every industry. As you plan your IoT career, consider not only the skills and roles that interest you, but also the industries and types of companies where you would like to work. This will help you align your career path with your interests and aspirations, and find a role where you can truly make an impact.

Trends in IoT Job Market

Understanding the trends in the IoT job market can help you stay ahead of the curve and seize exciting career opportunities. As a rapidly evolving field, IoT is continuously giving rise to new trends, shaping the demand for different skills, roles, and expertise. Let's explore some of the key trends in the IoT job market.

Rising Demand for IoT Skills

As more and more businesses and industries adopt IoT technologies, there is a growing demand for professionals with IoT skills. This demand spans a range of roles, from IoT developers and engineers to data analysts and project managers. Therefore, investing in building your IoT skills can significantly enhance your employability and career prospects.

Focus on Security

With the proliferation of IoT devices and the valuable data they generate, security has become a key concern. Consequently, there's an increasing demand for professionals who can ensure the security of IoT systems. This includes roles like IoT security analysts, who are responsible for identifying and mitigating security risks, and IoT security engineers, who design and implement security measures in IoT systems.

Edge Computing Skills

Edge computing, where data processing happens closer to the source of data (the "edge" of the network), is becoming increasingly important in IoT. This is due to the need for quicker response times and reduced bandwidth usage in many IoT applications. As a result, there's a growing demand for professionals with skills in edge computing. This could mean roles focused on designing and implementing edge computing infrastructure or developing IoT applications that leverage edge computing.

Data Analytics and AI

The vast amounts of data generated by IoT devices provide rich opportunities for analysis and insight generation. Consequently, there's a rising demand for professionals with skills in data analytics and AI, who can extract valuable insights from this data. This could include data scientists who analyze IoT data, machine learning engineers who build predictive models based on this data, or AI specialists who develop intelligent IoT applications.

Integration of IoT with Other Technologies

IoT doesn't exist in a vacuum - it's often integrated with other technologies, such as cloud computing, blockchain, and 5G. Understanding how to leverage these technologies in conjunction with IoT can open up new opportunities and make you more versatile as a professional. For example, you might work on a project that uses blockchain for secure data sharing between IoT devices, or develop an IoT application that leverages the speed and bandwidth of 5G networks.

Remember, staying updated on the trends in the IoT job market is crucial for staying relevant and competitive. Keep learning, keep evolving, and keep an eye on the market trends. This will help you spot new opportunities, plan your learning journey, and shape a successful career in the vibrant world of IoT.

Tips for a Successful IoT Career

Building a successful career in IoT requires a combination of technical knowledge, practical skills, and strategic thinking. As the IoT landscape continues to evolve, it's essential to stay agile and proactive in your career development. Here are some tips to help you navigate your IoT career journey successfully.

Develop a Strong Foundation

Having a solid understanding of the basic principles and technologies of IoT is crucial. This includes areas like networking, programming, data management, and cybersecurity. Pursuing relevant educational pathways, such as degrees, online courses, or professional certifications, can help you build this foundation.

Stay Updated with Latest Trends

The IoT field is evolving rapidly, with new technologies and applications emerging regularly. Make it a habit to stay updated with the latest trends, either by reading industry news, attending webinars, or participating in professional associations and communities. This will help you stay relevant and be aware of new opportunities.

Gain Hands-On Experience

Practical experience is invaluable in the IoT field. Look for opportunities to gain hands-on experience, whether it's through your job, internships, or personal projects. For example, you might build your own IoT device or develop an IoT application. This will not only enhance your skills but also give you a deeper understanding of how IoT technologies work in practice.

Focus on Security

Security is a key concern in the IoT field. Make sure to acquire a good understanding of IoT security principles and practices. This includes areas like secure IoT design, data encryption, and security testing. You might also consider pursuing a certification in IoT security to further demonstrate your expertise.

Master Data Analytics

IoT generates vast amounts of data, and knowing how to analyze this data is a valuable skill. Gain proficiency in data analytics tools and techniques, and consider learning about machine learning and AI, which are often used to extract insights from IoT data.

Learn to Communicate and Collaborate

Effective communication and collaboration are critical in the IoT field, which often involves working in cross-functional teams. Develop your soft skills, such as communication, teamwork, and problem-solving. This will help you work effectively with others and advance in your career.

Think Strategically

Understanding the strategic implications of IoT can make you more valuable as a professional. Learn to think about how IoT can create value, drive innovation, and transform businesses. This strategic mindset can help you contribute more effectively to your organization and open up opportunities for leadership roles.

Be Adaptable and Resilient

The IoT field is fast-paced and dynamic, and it's important to be adaptable and resilient. Be open to learning new things, taking on new challenges, and stepping out of your comfort zone. Don't be afraid to make mistakes - they're often the best learning opportunities.

Remember, building a successful career in IoT is a journey, not a destination. Stay curious, keep learning, and keep pushing yourself. With persistence and the right approach, you can build a rewarding and impactful career in the IoT field.

Module 10 Exam Questions

1. What is an example of a job role in IoT?

a) Web Developer

b) HR Manager

c) Real Estate Agent

d) IoT Solutions Architect

2. What is an important skill for working in the IoT field?

a) Cooking

b) Programming

c) Drawing

d) Singing

3. What is an example of a tech company that offers IoT platforms and services?

a) McDonald's

b) Microsoft

c) Nike

d) Gucci

4. What is one of the sectors where IoT is making a significant impact?

a) Fashion

b) Music

c) Healthcare

d) Literature

5. Which company uses IoT to optimize its logistics operations?

a) Coca-Cola

b) Adidas

c) Airbnb

d) DHL

6. Which technology is becoming increasingly important in IoT due to the need for quicker response times and reduced bandwidth usage?

a) Typewriters

b) Photocopiers

c) Televisions

d) Edge Computing

7. What type of professional is in demand due to the vast amounts of data generated by IoT devices?

a) Flight Attendant

b) Fitness Instructor

c) Data Analyst

d) Tour Guide

8. What is a key concern in the IoT field that is creating demand for specific professionals?

a) Gardening

b) Security

c) Painting

d) Sculpting

9. In what sector is IoT being used for applications such as predictive maintenance and supply chain optimization?

a) Entertainment

b) Sports

c) Retail

d) Manufacturing and Industrial

10. What type of IoT professional is responsible for identifying and mitigating security risks?

a) IoT Chef

b) IoT Painter

c) IoT Musician

d) IoT Security Analyst

11. What type of professional designs and implements security measures in IoT systems?

a) IoT Designer

b) IoT Architect

c) IoT Security Engineer

d) IoT Marketer

12. What is one area of focus when developing a strong foundation for an IoT career?

a) Dancing

b) Writing

c) Sculpting

d) Networking

13. What does a Machine Learning Engineer do in the IoT field?

a) Designs IoT devices

b) Manages HR for an IoT company

c) Writes articles about IoT

d) Builds predictive models based on IoT data

14. What is an example of a telecommunications company that is heavily involved in the IoT space?

a) Starbucks

b) Amazon

c) Walmart

d) Verizon

15. How can you gain practical experience in the IoT field?

a) By learning to play a musical instrument

b) By starting a book club

c) By creating a painting

d) By building your own IoT device

16. What is a valuable skill in the IoT field due to the vast amounts of data generated by IoT devices?

a) Singing

b) Cooking

c) Painting

d) Data Analytics

17. What is a key principle to consider when designing IoT systems?

a) Aesthetics

b) Humor

c) Artistry

d) Security

18. What is a necessary skill for working effectively in cross-functional teams in the IoT field?

a) Juggling

b) Magic Tricks

c) Communication

d) Sewing

19. What mindset can help you contribute more effectively to your organization and open up opportunities for leadership roles in the IoT field?

a) Athletic mindset

b) Artistic thinking

c) Strategic thinking

d) Culinary mindset

20. What is an important personal quality to have in the fast-paced and dynamic IoT field?

a) Inflexibility

b) Impatience

c) Resistance to change

d) Adaptability and resilience

Answers:

1. d
2. b
3. b
4. c
5. d
6. d
7. c
8. b
9. d
10. d
11. c
12. d
13. d
14. d
15. d
16. d
17. d
18. c
19. c
20. d

Conclusions and Next Steps

As we conclude this book, it's evident that the Internet of Things (IoT) revolution is reshaping our world, with its profound influence being felt across industries and in our personal lives. Throughout this exploration, we've examined how IoT is not just a passing trend, but a transformative force, driving economic growth, enhancing our standard of living, and revolutionizing industries. We've delved into strategies for robust IoT system management, mitigating potential security risks, and fostering innovation and leadership in IoT implementation.

The future will be defined by the pervasive integration of IoT, transforming industries and making our lives more connected than ever before. Understanding the capabilities and challenges of IoT will allow us to leverage this technology to create smarter, safer, and more efficient systems and communities. However, this transformation truly flourishes when we each actively embrace the changes brought about by the IoT era.

As a change-maker in this space, you have the capacity to significantly influence your industry and the wider world. Here are the next steps you can take to embrace the IoT revolution and drive positive change:

1. **Enroll the online program at TheBestMBA.org:** Certify your skills, gain in-depth knowledge, practical tools, and valuable insights from our comprehensive curriculum. Our program is designed to help you thrive in the IoT era and adapt to the ever-changing demands of the workforce.
2. **Engage in continuous learning:** Stay ahead of the curve by constantly updating your skills and knowledge. Embrace lifelong learning and take advantage of the vast resources available online, such as courses, webinars, podcasts, and articles.
3. **Expand your network:** Connect with industry leaders, experts, and fellow game changers through IoT-focused events, conferences, and workshops. Foster meaningful

relationships and collaborations that will propel you forward in your career.

4. **Seek mentorship and guidance:** Reach out to experienced professionals in the field of IoT and related industries to learn from their insights and experiences. Their wisdom can help you navigate the complexities of the IoT landscape and ensure your success.

5. **Volunteer or collaborate on IoT projects:** Gain hands-on experience and demonstrate your capabilities by working on real-world IoT projects. This not only enhances your skills but also showcases your value to potential employers and collaborators.

By taking these steps, you will be well on your way to becoming an influential game changer in the IoT era. Remember, the journey towards a sustainable and inclusive future begins with each individual's commitment to adapt, learn, and grow. Embrace the possibilities that IoT offers and let your unique human skills shine through as you create a better tomorrow.

The future is now. Are you ready to shape it?

Join the IIEC Internet of Things (IoT) Certification Program today at https://thebestmba.org and become a game changer.

I hope you enjoyed this journey together, and as I mentioned at the beginning of this book, I would love for you to share your experience with me after completing this program. Feel free to message me via LinkedIn.

https://www.linkedin.com/in/angelberniz

To your success,

Angel Berniz

Made in the USA
Monee, IL
04 November 2026

67384054R00184